FINITE
MATHEMATICS

FINITE MATHEMATICS

DAVID A. SPRECHER
University of California, Santa Barbara

Harper & Row, Publishers
New York, Evanston, San Francisco, London

Sponsoring Editor: George J. Telecki
Project Editor: Brenda Goldberg
Designer: Michel Craig
Production Supervisor: Will C. Jomarrón
Compositor: Monotype Composition Company, Inc.
Printer and Binder: Halliday Lithograph Corporation
Art Studio: Eric G. Hieber Associates Inc.

FINITE MATHEMATICS

Library of Congress Cataloging in Publication Data

Sprecher, David A.   1930–
    Finite mathematics.

    Includes index.
    1. Mathematics—1961–   I. Title.
QA39.2.S686   510   75-26890
ISBN 0-06-046391-0

To Devora, Lorrie, and Jeannie

CONTENTS

PREFACE

This text offers an intuitive introduction to finite mathematics, suitable for a one-semester or one-quarter course. Finite mathematics courses are often elected by students who seek a basic mathematics course, and my purpose has been to introduce at this level ideas and techniques which, in addition to their intrinsic interest and challenge, are of great importance in many academic and nonacademic areas.

The approach adopted for this text is to teach the subject through active student participation. The conceptual development and the associated problem-solving techniques are, therefore, integrated and treated as inseparable components of the text. The numerous worked-out examples are used to bridge the gap between theory and exercises. They show how theoretical considerations lead to practical tools, and they can be used as models for doing the exercises. The large number of graded exercises should be regarded as an integral part of the text, and a good number of these should be done as a matter of routine. In addition to enhancing the learning process, they will also serve as an excellent indicator of the degree of comprehension of the material.

This text begins where the high school leaves off. For many students, however, it may have been one or more years since their last mathematics course, and like other unused languages and skills, mathematics is easily forgotten. The text therefore begins with a concise review of the needed terminology and facts, together with sets of exercises for regaining facility in computations.

Because of variability in student interests and goals, the chapters have been kept self-contained as much as possible. This gives desired flexibility in omitting material or covering it out of sequence. The extensive cross-referencing will direct the reader in one section to the relevant material elsewhere in the book.

A very basic course can be made up of Chapters 0 (as needed), 1, and 3–6, with the possible omission of Sections 1.4, 3.3, 3.7, 4.4,

6.4, and 6.6. An alternative choice would be to include Chapter 7 but omit Chapter 2 and Sections 3.3, 3.7, and 6.6. Yet another choice for a short course could consist of Chapters 0–6, with the possible omission of Sections 2.5, 2.6, 3.3., and 6.6.

I am indebted to many individuals who, one way or another, extended their help and advice. I wish to express particular appreciation to Jean Mulder and Richard Hull for working the exercises and making useful editorial suggestions; to Sonia Ospina and Devora Sprecher for the expert typing of the manuscript; and to George Telecki, Brenda Goldberg, and the rest of the dedicated staff of Harper & Row for their support and attention during the different stages of the production of this book.

David A. Sprecher
Santa Barbara, California

FINITE
MATHEMATICS

O
REVIEW

0

The purpose of this material is to provide a quick review of the basic properties of rational numbers, powers, and inequalities. An inadequate understanding of these is a common cause of difficulties in computations, which often makes the difference between success and failure in a mathematics course. Working the exercises below is an effective way of brushing up on the basic manipulative skills required for this text, and many students should benefit from this.

0.1 THE REAL NUMBERS

We begin with an informal discussion of the real numbers for the purpose of recalling basic facts and terminology.

Most familiar to us are the *natural numbers* 1, 2, 3, ... to which we were introduced early in our education. These numbers, also called *counting numbers*, or *positive integers*, were then augmented to form the *integers* ..., $-3, -2, -1, 0, 1, 2, 3, \ldots$. Many problems, such as those involving measures and weights, cannot be solved with integers, and so the *rational numbers* were introduced. The most important discovery at this stage was the fact that numbers can be associated with points on a straight line. Let us briefly describe how this is done.

On a straight line, taken for convenience to be horizontal, we select an arbitrary point as the *origin* and label it 0; a second point is chosen to its right and labeled 1. The line segment $\overline{01}$ is called *unit length* (see Figure 0.1), and with it we mark equally spaced points to the right and to the left of 0. The point lying n units to the right of 0 is labeled n, and the point lying n units to the left of 0 is labeled $-n$.

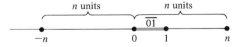

FIGURE 0.1

Dividing each interval so created into two equal parts gives the points associated with $\frac{1}{2}$, $-\frac{1}{2}$, $\frac{3}{2}$, $-\frac{3}{2}$, and so on. In general, for any positive integer n, the points associated with the numbers $1/n$, $-1/n$, $2/n$, $-2/n$, $3/n$, $-3/n$, and so on, are obtained by dividing each of the intervals created by the integers into n equal parts.

Of great significance in mathematics is the fact that infinitely many points on the line have no rational numbers corresponding to them. Numbers which cannot be written as a quotient of two integers are called *irrational*, and examples of such numbers are $\sqrt{2}$, $\sqrt{3}$, and π. Together, the rational and irrational numbers make up the *real numbers*.

Many examples and exercises in subsequent sections involve manipulations with rational numbers. For the purpose of this text, a rational number can be thought of as a fraction a/b, where a and b are integers and $b \neq 0$. We observe that $2/10$, $3/15$, and $6/30$, all represent the same rational number, which in *lowest terms* is $1/5$. Equality of rational numbers is defined as follows:

Equality of rational numbers

$$\frac{a}{b} = \frac{c}{d} \quad \text{if and only if} \quad ad = bc$$

Thus,

$$\frac{2}{10} = \frac{3}{15} \quad \text{since} \quad 2 \times 15 = 10 \times 3$$

and so on.

The arithmetical operations of rational numbers satisfy the following rules.

Rules for operating with rational numbers

(1) $\quad \dfrac{a}{b} + \dfrac{c}{d} = \dfrac{ad + bc}{bd}$

(2) $\quad \dfrac{a}{b} - \dfrac{c}{d} = \dfrac{ad - bc}{bd}$

(3) $\quad \dfrac{a}{b} \times \dfrac{c}{d} = \dfrac{a \times c}{b \times d}$

(4) $\quad \dfrac{a}{b} \Big/ \dfrac{c}{d} = \dfrac{a \times d}{b \times c}$

Example 1

(1) $\dfrac{2}{3} + \dfrac{5}{7} = \dfrac{2 \times 7 + 3 \times 5}{3 \times 7} = \dfrac{29}{21}$

(2) $\dfrac{2}{3} - \dfrac{5}{7} = \dfrac{2 \times 7 - 3 \times 5}{3 \times 7} = \dfrac{-1}{21} = -\dfrac{1}{21}$

(3) $\dfrac{2}{3} \times \dfrac{5}{7} = \dfrac{2 \times 5}{3 \times 7} = \dfrac{10}{21}$

(4) $\dfrac{2}{3} \Big/ \dfrac{5}{7} = \dfrac{2 \times 7}{3 \times 5} = \dfrac{14}{15}$

The following summation formula is often useful:

$$1 + 2 + 3 + \cdots + n = \frac{n(n + 1)}{2} \quad \text{for any positive integer } n$$

This formula is derived as follows: Suppose the sum of the first n positive integers equals A. Then we can write down the following scheme:

$$
\begin{array}{ccccccccc}
1 & + & 2 & + & 3 & + \cdots + & (n-1) & + & n & = A \\
n & + & (n-1) & + & (n-2) & + \cdots + & 2 & + & 1 & = A \\
\hline
(n+1) & + & (n+1) & + & (n+1) & + \cdots + & (n+1) & + & (n+1) & = 2A
\end{array}
$$

$$\underbrace{}_{n \text{ terms}}$$

From this we see that

$$2A = n(n + 1)$$

and hence

$$A = \frac{n(n + 1)}{2}$$

Example 2

$$1 + 2 + 3 + \cdots + 100 = \frac{100 \times 101}{2} = 5050$$

EXERCISES

1. Insert one of the symbols $=$ or $\neq$ into each box to produce a true statement.

(a) $1 \left/ \dfrac{a}{b} \right. \square \dfrac{b}{a}$

(b) $\dfrac{a}{b} - c \,\square\, \dfrac{bc}{a}$

(c) $\dfrac{2a + 3b}{4a + 9b} \,\square\, \dfrac{a + b}{2a + 3b}$

(d) $\dfrac{a + b - c}{b} \,\square\, 1 + \dfrac{a - c}{b}$

(e) $\dfrac{1}{b} \left/ 1 \right. \square\, b$

(f) $\dfrac{a}{b} \times \dfrac{c}{d} \times \dfrac{d}{a} \,\square\, \dfrac{d}{b}$

(g) $2 + 4 + 6 + 8 + 10 + 12 \,\square\, 2 \times \dfrac{6 \times 7}{2}$

(h) $5 + 10 + 15 + \cdots + 100 \,\square\, 5 \times \dfrac{20 \times 21}{2}$

Simplify the following fractions as much as possible.

2. $\dfrac{\frac{1}{2}}{2}$

3. $\dfrac{12}{-\frac{1}{3}}$

4. $\dfrac{\frac{13}{1}}{\frac{1}{13}}$

5. $\dfrac{\frac{1}{13}}{\frac{13}{1}}$

6. $\dfrac{-\frac{5}{2}}{\frac{3}{2}}$

7. $\dfrac{\frac{7}{2} - 1}{\frac{7}{2} + 1}$

8. $\dfrac{9 - \frac{1}{3}}{\frac{1}{3}}$

9. $\dfrac{-\frac{1}{4} + 2}{\frac{1}{4} - 2}$

10. $\dfrac{\frac{1}{2} - \frac{1}{3}}{\frac{1}{2} + \frac{1}{3}}$

11. $\dfrac{\frac{2}{3} - \frac{3}{5}}{\frac{3}{5} - \frac{2}{3}}$

12. $\dfrac{\frac{1}{2}}{\frac{1}{7} - \frac{1}{9}}$

13. $\frac{1}{2} + \frac{1}{3} - \frac{1}{7}$

14. $\dfrac{\frac{1}{2}}{\frac{1}{3}}$

15. $\frac{1}{2} + \frac{1}{4} + \frac{1}{8}$

16. $\dfrac{\dfrac{1}{a} - \dfrac{1}{b}}{\dfrac{1}{ab}}$ 17. $1 - \dfrac{a}{a+b}$

18. $\dfrac{4a + b}{a} - \dfrac{b}{a}$

0.2   INTEGRAL POWERS

Powers (exponents) play a very important role in computations. Here we require only integral powers, but powers a^x can be defined for any real number x.

DEFINITION OF INTEGRAL POWERS

(1)   If a is any real number, then

$$a^m = \underbrace{a \times a \times a \times \cdots \times a}_{m \text{ times}} \quad \text{for} \quad m = 1, 2, 3, \ldots$$

(2) $a^0 = 1$ for $a \neq 0$

(3) $a^{-m} = \dfrac{1}{a^m}$ for $a \neq 0$

Example 1

(1) $4^3 = 4 \times 4 \times 4 = 64$

(2) $4^0 = 1$

(3) $4^{-2} = \dfrac{1}{4^2} = \dfrac{1}{16}$

Properties of powers

Let m and n be any integers. Then

(1) $a^m a^n = a^{m+n}$

(2) $\dfrac{a^m}{a^n} = a^{m-n}$

(3) $(a^m)^n = a^{mn}$

(4) $(ab)^m = a^m b^m$

(5) $\left(\dfrac{a}{b}\right)^m = \dfrac{a^m}{b^m}$

Example 2

(1) $\quad 4^5\left(\dfrac{1}{4}\right)^5 = \left(4 \times \dfrac{1}{4}\right)^5 = 1^5 = 1$

(2) $\quad \left(\dfrac{2}{3}\right)^{-2}\left(\dfrac{3}{2}\right)^2 = \dfrac{1}{\left(\frac{2}{3}\right)^2}\left(\dfrac{3}{2}\right)^2 = \left(\dfrac{3}{2}\right)^2\left(\dfrac{3}{2}\right)^2 = \left(\dfrac{3}{2}\right)^4 = \dfrac{81}{16}$

(3) $\quad \dfrac{10^{-7} \times 10^{12}}{10^{-1}} = 10^{-7+12+1} = 10^6$

The following summation formula is often useful in computations:

$$1 + a + a^2 + a^3 + \cdots + a^n = \frac{1 - a^{n+1}}{1 - a} \qquad \text{whenever} \quad a \neq 1 \quad (1)$$

This formula is verified by multiplying both sides of the equation by $1 - a$, multiplying out $(1 - a)(1 + a + a^2 + a^3 + \cdots + a^n)$ and canceling.

Example 3

(1) $\quad 1 + \dfrac{1}{2} + \dfrac{1}{4} + \cdots + \dfrac{1}{64} = \dfrac{1 - \dfrac{1}{128}}{1 - \dfrac{1}{2}} = \dfrac{1 - \dfrac{1}{128}}{\dfrac{1}{2}}$

$$= 2\left(1 - \frac{1}{128}\right) = 2 - \frac{1}{64}$$

(2) $\quad 1 + 3 + 9 + 27 + 81 + 243 = \dfrac{1 - 729}{1 - 3} = \dfrac{-728}{-2} = 364$

EXERCISES

1. Insert one of the symbols $=$ or $\neq$ into each box to produce a true statement.

   (a) $\quad (-1)^{-1} \ \square \ 1$

   (b) $\quad \left(\dfrac{1}{a}\right)^{10} \ \square \ \dfrac{1}{a^{10}}$

   (c) $\quad a^5 a^{-5} \ \square \ 0$

   (d) $\quad (-1)^n + (-1)^{n+1} \ \square \ 0$

   (e) $\quad b^{-4}\left(\dfrac{a}{b}\right)^4 \ \square \ a^4 b^{-8}$

(f) $a \cdot a^2 \cdot a^3 \ \square \ a^6$

(g) $(a^{-2}b)^2 \ \square \ a^4b^2$

(h) $[(-2)^{-1}]^{-1} \ \square \ 2$

Simplify the numbers on Exercises 2–20 as much as possible, but leave
your answers expressed with positive powers.

2. $12\left(\dfrac{1}{2}\right)^3$

3. $\left(\dfrac{1}{3}\right)^3\left(\dfrac{2}{3}\right)$

4. $\left(\dfrac{4}{5}\right)\left(\dfrac{1}{5}\right)^4$

5. $\left(\dfrac{2}{3}\right)^4\left(\dfrac{1}{3}\right)^2$

6. $\left(\dfrac{5}{4}\right)^2\left(\dfrac{1}{5}\right)^3$

7. $\left(\dfrac{8}{9}\right)\left(\dfrac{1}{9}\right)^3$

8. $\left(\dfrac{9}{10}\right)^3\left(\dfrac{10}{12}\right)^{-1}$

9. $10^4 \times 10^{-7}$

10. $\dfrac{10^4 \times 10^{-6}}{10^{-4} \times 10^6}$

11. $(a^{-3})^{-3}$

12. $\left(\dfrac{a}{b} \Big/ \dfrac{c}{d}\right)^2$

13. $\dfrac{10^{-3}}{10^{-5}}$

14. $\dfrac{8^5}{2^{12}}$

15. $3^3 \times 3^{-4} \times 3^5$

16. $\left(\dfrac{1}{2}\right)^{-6}$

17. $\left(\dfrac{1}{5}\right)^4\left(\dfrac{2}{5}\right)^4$

18. $(2^2 \times 16)^8$

19. $(a^3b)(ab^3)$

20. $a^{2(m-1)}(a^{m-1})^2$

Find the sums in Exercises 21–25.

21. $1 + \dfrac{1}{2} + \cdots + \dfrac{1}{2^{10}}$

22. $1 + \dfrac{1}{5} + \cdots + \dfrac{1}{5^8}$

23. $1 + \dfrac{1}{10} + \cdots + \dfrac{1}{10^6}$

24. $1 - \dfrac{1}{2} + \dfrac{1}{4} - \dfrac{1}{8} + \cdots - \dfrac{1}{2^9}$

Hint:
Use Formula (1) with $a = -\frac{1}{2}$

25. $1 + \left(\dfrac{2}{3}\right) + \left(\dfrac{2}{3}\right)^2 + \left(\dfrac{2}{3}\right)^3 + \cdots + \left(\dfrac{2}{3}\right)^{10}$

0.3 INEQUALITIES

A basic property of real numbers is their order. This is expressed with the familiar symbols $>$ (greater than), $=$ (equal to), and $<$ (less than), and we use the assumption that any two numbers a and b are related by exactly one of the relations

$$a < b \qquad a = b \quad \text{or} \quad a > b$$

The symbols $\geq$ and $\leq$ are used for the following abbreviations:

$a \geq b$ is the same as "$a > b$ or $a = b$"
$a \leq b$ is the same as "$a < b$ or $a = b$"

Example 1
The statement "$4 < 4$ *or* $4 = 4$" is true because $4 = 4$. Hence, we can write $4 \leq 4$.
The statement "$-1 < 0$ *or* $-1 = 0$" is true because $-1 < 0$. Hence, we can write $-1 \leq 0$.

Terminology

$a > b$ is read "a is greater than b."
$a \geq b$ is read "a is greater than or equal to b."
$a < b$ is read "a is less than b."
$a \leq b$ is read "a is less than or equal to b."

It should be noted that $a < b$ and $b > a$ are two different ways of saying the same thing. Similarly, $a \leq b$ and $b \geq a$ convey the same information.

Properties of inequalities

(1)   If $a < b$ and $b < c$, then $a < c$.
(2)   If $a < b$, then $a + c < b + c$.
(3)   If $a < b$ and $c > 0$, then $ac < bc$.
(4)   If $a < b$ and $c < 0$, then $ac > bc$.

Example 2

(1) Since $2 < 3$, we have $-3 < -2$. Here we use property (4) with $c = -1$.
(2) Since $-5 < -4$ we have $-15 < -12$. Here we use property (3) with $c = 3$.

Example 3
Solve the inequality $4x - 1 < -x + 5$ for x.

SOLUTION

An inequality in an unknown x is solved like an equality. The following steps will illustrate the procedure:

$$4x - 1 < -x + 5 \quad \text{add } x \text{ to both sides}$$
$$5x - 1 < 5 \quad \text{add 1 to both sides}$$
$$5x < 6 \quad \text{divide both sides by 5}$$
$$x < \frac{6}{5}$$

Thus, the inequality $4x - 1 < -x + 5$ holds for all values of x less than $\frac{6}{5}$.

EXERCISES

1. Insert one of the symbols $<$ or $=$ or $>$ into each box to produce a true statement.

(a) $5^0 \ \square \ 0$

(b) $-1 \ \square \ -2$

(c) $-\dfrac{7}{15} \ \square \ -\dfrac{16}{30}$

(d) $0 - 1 \ \square \ 0 - \dfrac{1}{2}$

(e) $\dfrac{3}{5} \ \square \ \dfrac{11}{12}$

(f) $\left(-\dfrac{1}{3}\right)^4 \ \square \ \left(\dfrac{1}{3}\right)^3$

(g) $\frac{1}{2}/\frac{1}{3} \ \square \ 1$

(h) $\dfrac{1}{5} - \dfrac{1}{3} \ \square \ -1$

(i) $(2^{-1})^{-1} \ \square \ 2$

(j) $\left(\dfrac{1}{10}\right)^{-10}$ ☐ 10^{10}

(k) $2/\tfrac{1}{4}$ ☐ 1

2. Which of the following statements say the same as "$a < b$"?

    (a) $a - b < 0$
    (b) $a - b > 0$
    (c) $b - a > 0$
    (d) $-b < -a$
    (e) $5a < 5b$

3. Which of the following statements say the same as "$\dfrac{1}{a} \leq \dfrac{1}{b}$" when $a > 0$ and $b > 0$?

    (a) $b - a > 0$
    (b) $a - b \geq 0$
    (c) $b \leq a$
    (d) $b < a$
    (e) $4a \geq 4b$

4. Solve the following inequalities for x.

    (a) $2x + 1 > 0$
    (b) $-x - 2 < 0$
    (c) $5x - 3 > 3x + 1$
    (d) $-5x + 6 > 2x - 2$
    (e) $-2x + 7 < -4x + 5$

5. Below is a list of True-False questions. When a statement is false, you should give a *counterexample*. For instance, the statement "If $x < 1$ then $x > 0$" is false. A counterexample is "$x = -1$" because $-1 < 1$ is true, whereas $-1 > 0$ is false. It takes only one counterexample to make a statement false.

    (a) $15 \leq 15$
    (b) $2 \leq 4$
    (c) If $a < b$ then $a \leq b$
    (d) If $a < b$ then $b > a$
    (e) If $a + b > 0$ then $a > 0$
    (f) If $\dfrac{1}{a} < \dfrac{1}{b}$ then $a > b$
    (g) If $a = 4$ then $a \leq 5$
    (h) If $a < 0$ then $-a > 0$
    (i) If $a < b$ then $-a > -b$
    (j) If $a > b$ then $-a < b$

1
SETS

1

Few concepts are more basic in everyday life than the concept of a collection of objects. Whether our collection consists of books, clothes, coins, or numbers, we know how to recognize it, either through familiarity with its objects or through familiarity with the properties of the objects. A collection in mathematics and its applications is called a *set*, and objects are called *elements* or *members*. Describing sets in precise terms and formalizing their properties resulted in the *theory of sets* which is a creation of great elegance and utility. This theory, developed by Georg Cantor in the 1870s, provides a unifying language and an important tool for describing and studying phenomena in many disciplines. We shall touch upon only a few basic ideas of the theory of sets, but enough to give us an appreciation of this area of human thought by the end of this course.

1.1 SETS, THEIR UNION AND INTERSECTION

A set, in its intuitive meaning as a collection of objects, is completely determined by its members: If all members of a set are known, then the set itself is known. Conversely, for any set A we can always decide whether a given element belongs to A. This leads us to state the following definition:

Equality of sets

The sets A and B are *equal*, written as

$$A = B$$

when they have precisely the same members. When A and B are *not equal*, we write

$$A \neq B$$

Example 1
Consider the sets

$A = \{1, 2, 3, 4\}$
$B = \{2, 1, 4, 3\}$
$C = \{1, 2, 2, 3, 4\}$

Each of these sets consists of the elements 1, 2, 3, 4, and since a set is completely determined by its members, we have

$A = B = C$

In general, new sets *cannot* be formed from a given set by changing the order in which its members are listed, nor by repeating members more than once.

Example 2
Consider the set

$A = \{a, b, c\}$

From the elements of this set we can form other sets, such as

$\{a\}$ $\{b\}$ $\{c\}$ $\{a, b\}$ $\{b, c\}$ and $\{a, c\}$

These sets are all distinct and they are said to be *subsets* of A.

Subsets

A is a *subset* of B, expressed symbolically as

$A \subset B$ or $B \supset A$

if each member of A is a member of B. When A is *not* a subset of B, we write

$A \not\subset B$ or $B \not\supset A$

The concept of subset is clearly brought out in Figure 1.1, where sets are represented by areas. Such pictorial representations of sets are called *Venn diagrams*, and they are indispensable in handling many problems involving sets.

The symbolic statement $A \subset B$ is often read "A is contained in

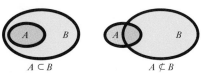

$$A \subset B \qquad A \not\subset B$$

FIGURE 1.1 *Subsets*

B," and $B \supset A$ is read "B contains A." When B has at least one member which is not in A, we call A a *proper subset* of B and write

$$A \subsetneq B \quad \text{or} \quad B \supsetneq A$$

The relation

$$A \subset A$$

is true for any set A.

Example 3
Let $A = \{1, 2, 3\}$. Which of the following relations is true?

$$1 \subset A, \quad \{1\} \subset A, \quad \{1, 2\} \subsetneq \{A\} \quad \text{and} \quad A \subsetneq A$$

SOLUTION

Since 1 is an element and not a set, writing $1 \subset A$ is meaningless. On the other hand, 1 is in A and hence $\{1\} \subset A$ is true. Since 3 is in A and not in $\{1, 2\}$ the relation $\{1, 2\} \subsetneq \{A\}$ is true, but the relation $A \subsetneq A$ is false because A is not a proper subset of itself.

Let us introduce set operations for building new sets from given ones.

Union of sets

The set of all members belonging to A *or* B is called the *union* of A and B, written

$$A \cup B$$

The concept of union is illustrated in Figure 1.2, where it is seen that the set $A \cup B$ may contain members of A and B. We therefore use here the word "or" in the *inclusive* sense, which is to say that "A or B" means "A or B or both A and B."

Example 4
Find $A \cup B$ when $A = \{0, 1, 2, 3\}$ and $B = \{2, 3, 4, 5, 6\}$.

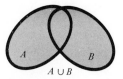

FIGURE 1.2   *Union of sets*

SOLUTION

The totality of elements to be found in A or B is 0, 1, 2, 3, 4, 5, 6. Hence

$$A \cup B = \{0, 1, 2, 3, 4, 5, 6\}$$

Notice specifically that the elements 2 and 3, which are members of both A and B, are not listed twice.

Intersection of sets

The set of all elements belonging to both A *and* B is called the *intersection* of A and B, written

$$A \cap B$$

The concept of intersection is illustrated in Figure 1.3, where $A \cap B$ is represented by the shaded area.

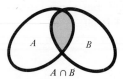

FIGURE 1.3   *Intersection of sets*

Example 5
Find $A \cap B$ when $A = \{0, 1, 2, 3\}$ and $B = \{2, 3, 4, 5, 6\}$.

SOLUTION

The elements in both A and B are 2 and 3. Consequently,

$$A \cap B = \{2, 3\}$$

In defining the intersection of sets we placed no restriction on the sets. To allow for the case when A and B have no elements in common, we introduce a set having no members. This set is designated with the symbol Ø and called the *empty set* or *null set*. It is assumed that

for any set A we have $\emptyset \subset A$

In words, the empty set is a subset of any set.

Example 6
Let A, B, and C be related as in Figure 1.4. Find the set

$$(A \cup B) \cap C$$

by crosshatching.

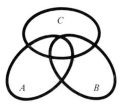

FIGURE 1.4

SOLUTION
Shading $A \cup B$ and C as in Figure 1.5, we obtain $(A \cup B) \cap C$ as the crosshatched area.

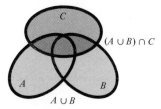

FIGURE 1.5

The operations of taking unions and intersections obey rules which are quite similar to the rules governing addition and multiplication of numbers. These rules are easily verified, using Venn diagrams and crosshatching.

Properties of union and intersection
If A, B, and C are sets then

(1) $A \cup B = B \cup A$
(2) $A \cup (B \cup C) = (A \cup B) \cup C$

(3) $A \cap B = B \cap A$
(4) $A \cap (B \cap C) = (A \cap B) \cap C$
(5) $A \cup (B \cap C) = (A \cup B) \cap (A \cup C)$
(6) $A \cap (B \cup C) = (A \cap B) \cup (A \cap C)$
(7) $A \cup A = A$
(8) $A \cap A = A$
(9) $A \cup \emptyset = A$
(10) $A \cap \emptyset = \emptyset$

EXERCISES

Find the sets in Exercises 1–9 when

$$A = \{1, 2, 3, 4\} \qquad B = \{5, 6\} \qquad C = \{3, 4, 5\}$$

1. $A \cup C$ 2. $(A \cup C) \cap B$
3. $A \cap B$ 4. $B \cap C$
5. $A \cup \emptyset$ 6. $A \cup B \cup C$
7. $A \cap B \cap C$ 8. $A \cap A$
9. $(A \cap C) \cup (B \cap C)$

For each of Exercises 10–16 draw a Venn diagram similar to Figure 1.6 and find the sets by crosshatching.

10. $(A \cup C) \cap B$ 11. $A \cup (B \cap C)$
12. $A \cap B \cap C$ 13. $D \cap (B \cup C)$
14. $B \cup (C \cap D)$ 15. $B \cup C \cup D$
16. $(A \cap B) \cup (C \cap D)$

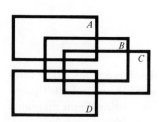

FIGURE 1.6   *Sample Venn diagram*

In Exercises 17–22 use Venn diagrams to show how the given sets must be related for the relations to be true.

17. $A \cap B = A$ 18. $A \cup B = A$

19. $(A \cup B) \cap C = A \cup B$ 20. $(A \cup B) \cap C = \emptyset$
21. $C \subset A \cap B$ 22. $C \supset B$ and $B \supset A$

23. Explain why statements (1) and (2) below say the same thing.

    (1) $A = B$
    (2) $A \subset B$ and $B \subset A$

24. Let $A = \{0, 1\}$. Which of the following relations is true?

    $A \subset \{0, 1\}$ $0 \subset A$ $\{0\} \subset A$

25. Let $A = \{\emptyset\}$. Which of the following relations is true?

    $A = \emptyset$ $\emptyset \subset A$ $\{\emptyset\} \subset A$ $A \subset \{\emptyset\}$

26. Which of the following sets are the same as $\{1, 2\}$?

    (a) $\{1, 1, 2\}$
    (b) $\{1, \{1\}, 2\}$
    (c) $\{\{1\}, \{2\}\}$
    (d) $\{1\} \cup \{2\}$
    (e) $\{1, 2\} \cup \{2\}$

27. What is the relation between the elements a and b if $\{a\} = \{a, b\}$?
28. What is the relation between the elements a, b, and c if

    $\{a, b\} \supset \{a, b, c\}$ and $\{a\} \cap \{c\} = \emptyset$?

29. The possible subsets of $\{0, 1\}$ are $\emptyset$, $\{0\}$, $\{1\}$, and $\{1, 2\}$. List all possible subsets of $\{0, 1, 2\}$.
30. What are the possible subsets of $\{a, b\}$? Of $\{a, b, c\}$?
31. When a coin is tossed twice, the set of possible outcomes is

    $S = \{HH, HT, TH, TT\}$

    where H = Head and T = Tail. List the set of possible outcomes when a coin is tossed three times, and write out the following subsets of possible outcomes.

    (a) more than one Head
    (b) exactly one Head
    (c) no tails

32. A die is rolled. What is the set of possible outcomes?

33. Two balls are drawn from an urn containing one white, one red, and one black ball. Write out the set of different ways in which this can be done if

(a) The two balls are drawn simultaneously.
(b) First one ball is drawn, replaced, then a second ball is drawn.

34. Consider the sets $A = \{a_1, a_2\}$ and $B = \{b_1, b_2, b_3\}$. Write out the set S of all possible sets formed from one element of A and one element of B.

35. Consider the sets $A = \{a_1, a_2\}$, $B = \{b_1, b_2\}$, and $C = \{c_1, c_2\}$. Write out the set S consisting of all sets formed from one element each from A, B, and C.

1.2 COMPLEMENT AND RELATIVE COMPLEMENT

We spoke about the elements in a given set A, but what about the elements *not* in A? Suppose, for example, that F is the set of female United States residents. When we speak of elements not in F, we usually refer to other humans, such as male residents of the United States. In general, in any given discussion we assume that all elements belong to some set, called *universal set*. The letter $\mathcal{U}$ will be used to designate such a set; in Venn diagrams it is customarily represented by the points of a rectangle (see Figure 1.7). Thus, in any given discussion, all sets are subsets of some universal set. The universal set is often not specified.

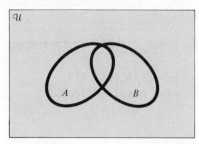

FIGURE 1.7   *A Universal set* $\mathcal{U}$ *with subsets* A *and* B

Complement of a set

The set of all elements of $\mathcal{U}$ which are *not* in A is called the *complement* of A, designated $\overline{A}$.

In agreement with the above remarks we have

$$A \cup \overline{A} = \mathcal{U}$$

(See Figure 1.8.)

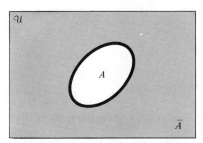

FIGURE 1.8   *The complement of* A

Example 1
List all subsets of the universal set $\mathfrak{U} = \{T_1, T_2, T_3\}$.

SOLUTION
The subsets of $\mathfrak{U}$ are

$$\emptyset, \quad \{T_1\}, \quad \{T_2\}, \quad \{T_3\}, \quad \{T_1, T_2\}, \quad \{T_1, T_3\}, \quad \{T_2, T_3\}, \mathfrak{U}$$

It is easy to see that we exhausted all possibilities. Observe that the set $\mathfrak{U}$ has three elements and $8 = 2^3$ subsets. We shall develop later a formula showing that a set of n elements has 2^n subsets.

Example 2
Let $\mathfrak{U} = \{T_1, T_2, T_3\}$, $A = \{T_1\}$, $B = \{T_2, T_3\}$. Find $\overline{A}$, $\overline{B}$, $\overline{A \cup B}$, and $\overline{A \cap B}$.

SOLUTION
We see at once that

$$\overline{A} = B \quad \text{and} \quad \overline{B} = A$$

Since $A \cup B = \mathfrak{U}$, we find that there are no elements not in $A \cup B$. Hence,

$$\overline{A \cup B} = \emptyset$$

Since $A \cap B = \emptyset$, we find that the elements *not* in $A \cap B$ are T_1, T_2, and T_3. Hence,

$$\overline{A \cap B} = \mathfrak{U}$$

Relative complement of a set
The set of all elements in A and not in B is called the *complement of* B *relative to* A, designated

$$A - B$$

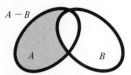

FIGURE 1.9 *Complement of* B *relative to* A

and read "*A* minus *B*." The concept of relative complement is illustrated in Figure 1.9.

Example 3
Verify the relation

$$A - B = A \cap \overline{B}$$

using Venn diagrams.

SOLUTION

Consult Figure 1.10. The set *A* is shaded in the left diagram and the set $\overline{B}$ is shaded in the middle diagram. Superimposing these two, we get the set $A \cap \overline{B}$ as the crosshatched area in the right figure. This, however, is exactly $A - B$.

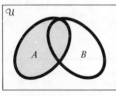

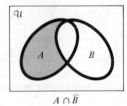

FIGURE 1.10

Example 4
Consider the average income tax deductions for 1968 listed in Table 1.1.

Let the letters below stand for deductions as follows:

L	low income	*X*	interest and taxes
M	medium income	*Y*	taxes and medical
H	high income	*Z*	contributions

Find the total deductions in the following sets:

(1) $L \cap X$
(2) $(L \cup M) \cap (X - Y)$
(3) $H \cap \overline{Z}$

TABLE 1.1 AVERAGE INCOME TAX DEDUCTIONS IN DOLLARS (1968)

	Contributions	Interest	Taxes	Medical
Low income	200	385	340	375
Medium income	325	730	710	415
High income	2,265	2,815	3,515	2,085

SOLUTION

(1) The set $L \cap X$ consists of the intersection of the low-income row with the interest and taxes columns as indicated in the first diagram below. The answer is therefore the sum of numbers in the crosshatched area. This sum is

$$385 + 340 = 725 \text{ dollars}$$

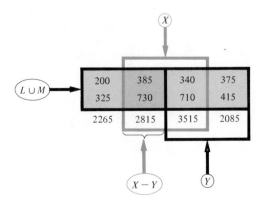

(2) The set $L \cup M$ consists of the low- and medium-income rows, and $X - Y$ stands for the interest column, as is evident from the next diagram below. The total deductions in $(L \cup M) \cap (X - Y)$ is therefore the sum of the numbers in the crosshatched area:

$$385 + 730 = 1,115 \text{ dollars}$$

(3)   Here the answer is

$$2,815 + 3,515 + 2,085 = 8,415 \text{ dollars}$$

The reason is that H represents the last row and $\bar{Z}$ stands for all deductions which are not contributions.

The operations of taking complements and relative complements of sets obey the following rules, which are easily verified with Venn diagrams.

Properties of complements and relative complements

Let A and B be any subsets of a universal set $\mathfrak{U}$. Then

(1) $\overline{(\overline{A})} = A$
(2) $A \cup B = \overline{(\overline{A} \cap \overline{B})}$
(3) $A \cap B = \overline{(\overline{A} \cup \overline{B})}$
(4) $\overline{A \cup B} = \overline{A} \cap \overline{B}$
(5) $\overline{A \cap B} = \overline{A} \cup \overline{B}$
(6) $A \cup \overline{A} = \mathfrak{U}$
(7) $A \cap \overline{A} = \emptyset$
(8) $A - A = \emptyset$
(9) $A - B = A \cap \overline{B}$
(10) $A - \emptyset = A$

EXERCISES

Find the sets in Exercises 1–12 when

$$\mathfrak{U} = \{1, 2, 3, 4, 5, 6\}$$
$$A = \{1, 2, 3, 4\}$$
$$B = \{5, 6\}$$
$$C = \{3, 4, 5\}$$

1. $A - C$ 2. $C - A$
3. $\overline{C - A}$ 4. $C - \overline{A}$
5. $\overline{B}$ 6. $\overline{B \cap C}$
7. $\overline{A} \cap \overline{B}$ 8. $B \cup \overline{B}$
9. $\overline{B \cup \overline{C}}$ 10. $\overline{A} \cap B$
11. $\overline{C} \cap B$ 12. $A \cap \mathfrak{U}$

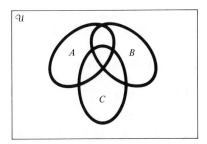

FIGURE 1.11 *Sample Venn diagram*

For each of Exercises 13–26 draw a Venn diagram similar to Figure 1.11 and find the sets by crosshatching.

13. $\overline{A} \cap \overline{B}$

14. $\overline{A \cap B}$

15. $\overline{A \cup B}$

16. $\overline{A} \cup \overline{B}$

17. $\overline{(B \cap C)}$

18. $(A - B) \cup (B - A)$

19. $A - \overline{B}$

20. $(B \cup C) \cap \overline{B}$

21. $\overline{A \cup B \cup C}$

22. $\overline{A - B} \cap \overline{C}$

23. $\overline{A} \cap \overline{B} \cap \overline{C}$

24. $\overline{(\overline{A})}$

25. $\overline{A} \cup (B \cap C)$

26. $A - (B \cup C)$

Exercises 27–34 refer to Table 1.2.

TABLE 1.2 FEDERAL TAX COLLECTIONS IN MILLIONS OF DOLLARS

Source \ Year	1929	1939	1949	1959
Excise taxes	540	1,768	7,579	10,760
Alcoholic beverages	13	588	2,211	3,002
Tobacco	434	580	1,322	1,807
Customs	602	319	384	948

Let revenues be designated with letters as follows:

E excise taxes
A alcoholic beverages
T tobacco
Q alcoholic beverages and tobacco
X 1929 plus 1939 revenues

Y 1949 revenues

$\mathfrak{u}$ total revenues for the four years listed in Table 1.2.

Find the amount of revenue in each of the following sets:

27. $E \cap X$ 28. $E \cap \mathfrak{u}$
29. $\overline{Q} \cap \overline{(X \cup Y)}$ 30. $\overline{(E \cup Q)} \cap X$
31. $\overline{E \cup Q} \cap \overline{(X \cup Y)}$ 32. $(E - A) \cap (X - Y)$
33. $\mathfrak{u} - (Q \cap Y)$ 34. $\overline{(E \cap Y)}$

35. Find all subsets of $\mathfrak{u} = \{a, b, c\}$.
36. Find all subsets of $\mathfrak{u} = \{a, b, c, d\}$.

Exercises 37–44 refer to the following concept: The *symmetric difference* of sets A and B is defined to be

$$A \triangle B = (A - B) \cup (B - A)$$

Find each of the following sets when

$$A = \{1, 2, 3\} \qquad B = \{2, 3, 4\} \qquad \mathfrak{u} = \{0, 1, 2, 3, 4, 5, 6\}$$

37. $A \triangle B$ 38. $(A \cap \overline{B}) \cup (\overline{A} \cap B)$
39. $A \triangle \overline{B}$ 40. $A \triangle A$
41. $A \triangle \emptyset$ 42. $A \cap (B \triangle \overline{B})$
43. $(A \cap \overline{A}) \triangle (B \cap \overline{B})$ 44. $\overline{(A \triangle B)}$

Express the sets in Exercises 45–50 symbolically, using the set operations.

45.   46.

47.   48.

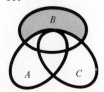   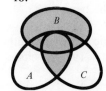

49. 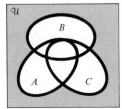   50.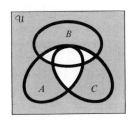

1.3 CONDITIONS

Up to now we specified sets by listing their members. This is not always feasible because a given set may be too large and we have to resort to a different method. There is another reason for being dissatisfied with our description of sets. It does not bring into play any of the properties which distinguish the elements of a given set from those not in the set. What we are after, then, is a description of sets which involves conditions on their members. We begin by introducing the following notation.

Membership in a set

If A is a given set and x an element, then the designation

$$x \in A$$

reads "x is a member of A" or "x belongs to A." The designation

$$x \notin A$$

reads "x is not a member of A" or "x does not belong to A." This definition is illustrated in Figure 1.12.

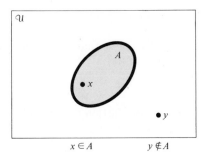

FIGURE 1.12 *Membership in a set*

Example 1

Let the universal set $\mathcal{U}$ consist of the natural numbers

$$\mathcal{U} = \{1, 2, 3, 4, \ldots\}$$

the three dots indicating that the listing goes on and on, without ever terminating. The statement "x is even" is a *condition* on x because each x in $\mathfrak{U}$ converts this statement into a true or a false statement:

$x = 1$ produces the *false* statement "1 is even"
$x = 2$ produces the *true* statement "2 is even"

Thus, the set of even natural numbers can be specified symbolically as

$$\{x \in \mathfrak{U} | x \text{ is even}\}$$

read "the set of all x in $\mathfrak{U}$ such that x is even." We often omit reference to $\mathfrak{U}$ and simply write

$$\{x | x \text{ is even}\} \tag{1}$$

Let us generalize the discussion in Example 1. Let $p(x)$ stand for the statement "x is even." Then $p(1)$ is a false statement and $p(2)$ is a true statement. The set in (1) can evidently be written as

$$\{x | p(x)\}$$

read "the set of all x such that p of x is true." In what follows, we shall let $p(x)$, $q(x)$, and so on, stand for statements about x. A statement $p(x)$ may be either true or false for a given x.

Example 2

(1) Let $p(x)$ represent the condition "$x = 0$ or $x = 1$." Then $p(0)$ and $p(1)$ are true statements, but for any other number x, $p(x)$ is false.
(2) Let $q(x)$ represent the condition "x is less than 10." Then $q(5)$ is a true statement but $q(10)$ is false.

Truth sets

Let $\mathfrak{U}$ be a given universal set and let $p(x)$ be a condition on the members $x \in \mathfrak{U}$. Then the set of all x in $\mathfrak{U}$ for which $p(x)$ is a true statement is designated

$$\{x | p(x)\} \quad \text{or} \quad \{x \in \mathfrak{U} | p(x)\}$$

and called the *truth set* of $p(x)$. The meaning of truth set is graphically explained in Figure 1.13.

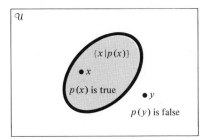

FIGURE 1.13   *The truth set of* p(x)

Example 3

(1)   Let $p(x)$ be the condition of being 0 or 1. Then

$$\{x|p(x)\} = \{x|x = 0 \quad \text{or} \quad x = 1\} = \{0, 1\}$$

(2)   Let $p(x)$ be the condition of *not* being a member of A. Then

$$\{x|p(x)\} = \{x|x \notin A\} = \overline{A}$$

(3)   Let $p(x)$ be the condition of being a number not less than 0 and not greater than 1. Then

$$\{x|p(x)\} = \{x|0 \leq x \leq 1\}$$

We can thus use conditions to specify sets. The relation between conditions and sets is not one-sided, however, and later we shall see how to use sets to analyze statements. We alluded to this possibility by calling $\{x|p(x)\}$ the *truth set* of the condition statement $p(x)$.

Using the condition as a set builder, we can give the following formal definition of the basic set operations.

DEFINITION 1

$$A \cup B = \{x|x \in A \quad \text{or} \quad x \in B\}$$
$$A \cap B = \{x|x \in A \quad \text{and} \quad x \in B\}$$
$$A - B = \{x|x \in A \quad \text{and} \quad x \notin B\}$$
$$\overline{A} = \{x|x \notin A\}$$

Example 4
Let

$$\mathcal{U} = \{x|x = \text{United States citizen}\}$$
$$M = \{x|x = \text{male}\}$$

$$F = \{x|x = \text{female}\}$$
$$W = \{x|x = \text{married}\}$$

Find

   (1) $M \cap W$
   (2) $\overline{M}$
   (3) $F - W$

SOLUTION

   (1) Using Definition 1 with $A = M$ and $B = W$ gives

$$
\begin{aligned}
M \cap W &= \{x|x \in M \quad \text{and} \quad x \in W\} \\
&= \{x|x = \text{male} \quad \text{and} \quad x = \text{married}\} \\
&= \{x|x = \text{married male}\}
\end{aligned}
$$

   (2) Our set $\mathfrak{U}$ consists of all males and females who are United States citizens. Hence,

$$\overline{M} = \{x|x \notin M\} = \{x|x \neq \text{male}\} = \{x|x = \text{female}\} = F$$

   (3) $
\begin{aligned}
F - W &= \{x|x \in F \quad \text{and} \quad x \notin W\} \\
&= \{x|x = \text{female} \quad \text{and} \quad x \neq \text{married}\} \\
&= \{x|x = \text{unmarried female}\}
\end{aligned}
$

Example 5

Referring to Table 1.3, let

$$\mathfrak{U} = \{1965, 1966, 1967, 1968\}$$

and consider the following conditions on the elements of $\mathfrak{U}$:

   $a(x)$: aluminum production exceeds 3,000,000 tons
   $c(x)$: copper production exceeds   1,300,000 tons
   $\ell(x)$: lead production exceeds   320,000 tons

Find

   (1) $\{x|a(x)\} \cup \{x|\ell(x)\}$
   (2) $\{x|a(x)\} \cap \{x|c(x)\}$
   (3) $\{x|a(x)\} - \{x|\ell(x)\}$
   (4) $\overline{\{x|a(x)\} \cap \{x|\ell(x)\}}$

TABLE 1.3 UNITED STATES METALS PRODUCTION (IN TONS)

Year (x)	Aluminum	Copper	Lead
1965	2,754,500	1,335,700	301,000
1966	2,968,400	1,353,100	327,400
1967	3,269,300	846,600	316,900
1968	3,255,000	1,161,000	354,200

SOLUTION

It is easy to see that

$$\{x|a(x)\} = \{1967, 1968\}$$
$$\{x|c(x)\} = \{1965, 1966\}$$
$$\{x|\ell(x)\} = \{1966, 1968\}$$

and we find that

(1) $\{x|a(x)\} \cup \{x|\ell(x)\} = \{1967, 1968\} \cup \{1966, 1968\}$
$$= \{1966, 1967, 1968\}$$

(2) $\{x|a(x)\} \cap \{x|c(x)\} = \{1967, 1968\} \cap \{1965, 1966\} = \emptyset$

(3) $\{x|a(x)\} - \{x|\ell(x)\} = \{1967, 1968\} - \{1966, 1968\}$
$$= \{1967\}$$

(4) $\overline{\{x|a(x)\} \cap \{x|\ell(x)\}} = \overline{\{1968\}} = \{1965, 1966, 1967\}$

EXERCISES

Express the sets in Exercises 1–18 by listing their members. The universal set is $\mathfrak{U} = \{1, 2, 3, 4, \ldots\}$.

1. $\{x|(x-1)(x-2) = 0\}$ **Answer** $\{1, 2\}$
2. $\{x|x < 5\}$
3. $\{x|x < 5.2\}$
4. $\{x|x < 1\}$
5. $\{x|x = -x\}$
6. $\{x|x \neq x\}$
7. $\{x|x^2 - 5x + 6 = 0\}$

8. $\{x|x$ is odd$\}$
9. $\{x|x = 4k$ and $k = 1, 2, 3, 4, \ldots\}$
10. $\overline{\{x|x > 5\}}$
11. $\overline{\{x|x \neq 1\}}$
12. $\{x|x \leq 10\} \cap \{x|x \geq 10\}$
13. $\{x|x = -x\} \cap \{x|x \geq 10\}$
14. $\{x|1 < x < 5\} \cup \{x|3 < x < 8\}$
15. $\{x|x < 10\} - \{x|x < 5\}$
16. $\{x|x < 10\} - \{x|x > 5\}$
17. $\overline{\{x|x < 1\}}$
18. $\{x|x^2 = 1\} \cup \{x|x^2 = 2\}$

Write the sets in Exercises 19–23 in the form $\{x|p(x)\}$. The universal set $\mathfrak{U}$ is the set of real numbers.

19. All numbers greater than 0 and less than 1.

Answer $\{x|0 < x < 1\}$

20. All negative numbers.
21. All numbers not less than 10.
22. 5, 10, 15, 20, 25, $\ldots$, $5n$, $\ldots$
23. 2, 2^2, 2^3, 2^4, 2^5, $\ldots$, 2^n, $\ldots$

For each of the Exercises 24–30 draw a Venn diagram similar to Figure 1.14 and find the sets by crosshatching.

24. $\{x|x \in A$ and $x \notin B\}$
25. $\{x|x \in \overline{A}\} \cup \{x|x \in B\}$
26. $\overline{\{x|x \in A\}} \cap \{x|x \in B\}$
27. $\{x|x \in \overline{A} - B\} \cap \{x|x \in C\}$
28. $\{x|x \in \overline{A} - B\} \cup \{x|x \in \overline{B} - A\}$
29. $\{x|x \in \overline{A \cap B}\} \cap \{x|x \in A \cap C\}$
30. $\{x|x \notin A\} \cup \{x|x \notin B\}$

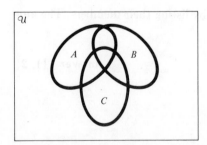

FIGURE 1.14   *Sample Venn diagram*

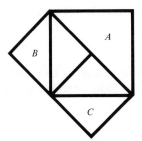

FIGURE 1.15

Consider the squares A, B, and C in Figure 1.15. Draw the following sets:

31. $\{x | x \in A$ and $x \in B\}$ 32. $\{x | x \in A$ or $x \in B\}$
33. $\{x | x \in A$ and $x \in C\}$ 34. $\{x | x \in A$ and $x \notin B\}$
35. $\{x | x \in A - C\}$ 36. $\{x | x \in A \cup C\} - \{x | x \in B\}$

1.4   EQUIVALENT CONDITIONS AND IMPLICATIONS

In the preceding section we saw how to build sets using conditions. We have not shown, however, how to use conditions to decide when two sets are equal or when one is a subset of the other.

Consider a universal set $\mathcal{U}$, conditions $p(x)$ and $q(x)$ on its members, and their truth sets, $\{x | p(x)\}$ and $\{x | q(x)\}$. If these sets are equal, then either $p(x)$ and $q(x)$ are both true or $p(x)$ and $q(x)$ are both false.

Equivalent conditions
If

$$\{x | p(x)\} = \{x | q(x)\}$$

then we say that $p(x)$ and $q(x)$ are *equivalent*, expressed

$$p(x) \Leftrightarrow q(x)$$

Thus, equivalent conditions are true at exactly the same elements of $\mathcal{U}$, and it follows that their truth sets are equal. Hence, we can state the following:

$$p(x) \Leftrightarrow q(x) \quad \text{if and only if} \quad \{x | p(x)\} = \{x | q(x)\}$$

The statement "X if and only if Y" means that either X and Y are both true, or they are both false.

Example 1
Let $\mathcal{U}$ consist of the points x in the plane and let C be a circle with center 0 and radius 1 (see Figure 1.16). Consider conditions $p(x)$ and $q(x)$ as follows:

$p(x)$ means that x lies inside the circle.
$q(x)$ means that the line segment joining x to 0 has length less than 1.

Then any point inside the circle has distance less than 1 from 0, and each point with distance less than 1 from 0 lies inside the circle. This implies that

$$\{x|p(x)\} = \{x|q(x)\}$$

and hence

$$p(x) \Leftrightarrow q(x)$$

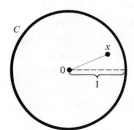

FIGURE 1.16   *A circle with center 0 and radius 1*

Example 2
Let $\mathcal{U}$ be the set of real numbers. Show that

$$x^2 - 1 = 0 \Leftrightarrow x^4 - 1 = 0$$

SOLUTION
We use the factorizations

$$x^2 - 1 = (x - 1)(x + 1)$$

and

$$x^4 - 1 = (x - 1)(x + 1)(x^2 + 1)$$

These can be verified by multiplying out the right side and simplifying. Now,

$$\{x|x^2 - 1 = 0\} = \{x|(x - 1)(x + 1) = 0\}$$
$$= \{x|x = 1 \quad \text{or} \quad x = -1\} = \{1, -1\}$$
$$\{x|x^4 - 1 = 0\} = \{x|(x - 1)(x + 1)(x^2 + 1) = 0\}$$

and since $x^2 + 1 = 0$ has no roots in $\mathfrak{u}$, it follows that

$$\{x|(x - 1)(x + 1)(x^2 + 1) = 0\}$$
$$= \{x|x = 1 \quad \text{or} \quad x = -1\} = \{1, -1\}$$

Hence,

$$\{x|x^2 - 1 = 0\} = \{x|x^4 - 1 = 0\}$$

and we have

$$x^2 - 1 = 0 \Leftrightarrow x^4 - 1 = 0$$

Consider once more conditions $p(x)$ and $q(x)$ with truth sets

$$P = \{x|p(x)\} \quad \text{and} \quad Q = \{x|q(x)\}$$

If $P \subset Q$ and $x \in P$, then

  (1) $p(x)$ is true.
  (2) $x \in Q$ so that also $q(x)$ is true.

In other words, the truth of $p(x)$ *implies* the truth of $q(x)$.

IMPLICATION
   If

$$\{x|p(x)\} \subset \{x|q(x)\}$$

then we say that $p(x)$ *implies* $q(x)$ and write

$$p(x) \rightarrow q(x)$$

Verbally stated, this definition says that if the truth set of $p(x)$ is contained in the truth set of $q(x)$, then $p(x)$ implies $q(x)$. Conversely, if

$p(x)$ implies $q(x)$, then for each x in P we have $x \in Q$ and hence $P \subset Q$. We summarize this as follows:

$$p(x) \rightarrow q(x) \quad \text{if and only if} \quad \{x|p(x)\} \subset \{x|q(x)\}$$

Remark

If

$$P = \{x|p(x)\} \quad \text{and} \quad Q = \{x|q(x)\}$$

and

$$P = Q$$

then

(1) $P \subset Q$ so that $p(x) \rightarrow q(x)$
(2) $Q \subset P$ so that $q(x) \rightarrow p(x)$

To show that sets are equal is most often done by showing that each is a subset of the other. Similarly, to show that two conditions are equivalent, one usually shows that each implies the other. It is thus useful to keep in mind the following facts:

$P = Q$ is the same as $P \subset Q$ and $Q \subset P$.
$p \Leftrightarrow q$ is the same as $p \rightarrow q$ and $q \rightarrow p$.

Comparing the symbols $\rightarrow$ and $\Leftrightarrow$ we note that

---

$\rightarrow$ stands for "if ... then."
$\Leftrightarrow$ stands for "if and only if."

Example 3
Let

$$\mathfrak{U} = \{x|x = \text{United States resident}\}$$

and consider the conditions $f(x)$ and $m(x)$ where

> $f(x)$ means x = female
> $m(x)$ means x = married female

Since every married female is a female, we have

$$\{x|m(x)\} \subset \{x|f(x)\}$$

and accordingly

$$m(x) \rightarrow f(x)$$

Example 4
Let $\mathfrak{U}$ be the set of real numbers. Is the implication

$$x > 5 \rightarrow x > 0$$

correct?

SOLUTION
To solve this problem, we merely show that

$$\{x|x > 5\} \subset \{x|x > 0\}$$

This, however, is clear from Figure 1.17. Observe that

$$\{x|x > 0\} \not\subset \{x|x > 5\}$$

and hence

$$x > 0 \not\rightarrow x > 5$$

The notation $p(x) \not\rightarrow q(x)$ means that $p(x)$ does *not* imply $q(x)$.

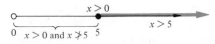

FIGURE 1.17

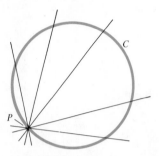

FIGURE 1.18   *The condition* p(x)

Example 5

Consider a circle C in the plane and a point P on it. Let $\mathfrak{U}$ consist of all lines in the plane and consider conditions $p(x)$ and $q(x)$ as follows:

$p(x)$ means that the line x passes through P (Figure 1.18).
$q(x)$ means that the line x meets C in two points (Figure 1.19).

Is the implication

$$p(x) \rightarrow q(x)$$

correct?

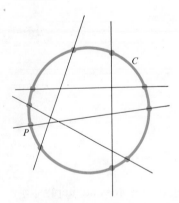

FIGURE 1.19   *The condition* q(x)

SOLUTION

For the implication to be valid, we must have

$$\{x|p(x)\} \subset \{x|q(x)\}$$

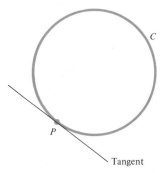

Tangent

FIGURE 1.20   *The tangent to the circle* C *at* P

The set $\{x|p(x)\}$, however, contains one line not in $\{x|q(x)\}$. This is the line which is *tangent* to the circle at P (see Figure 1.20). Hence,

$$\{x|p(x)\} \not\subset \{x|q(x)\}$$

and accordingly

$$p(x) \not\rightarrow q(x)$$

Example 6

In a class of ten students the following test scores were recorded:

Test A (x)	75	85	65	90	80	65	100	70	80	75
Test B (y)	60	90	75	85	90	80	95	80	95	90

Thus, for each student there is a pair of grades (x, y).

Write out the following sets:

(1) $\{(x, y)|x < 80 \rightarrow y > 80\}$
(2) $\{(x, y)|x \geq 75 \rightarrow y > 90\}$
(3) $\{(x, y)|x < 80 \Leftrightarrow y < 80\}$

SOLUTION

(1)   We observe that $x < 80 \rightarrow y > 80$ when $x < 80$ and $y > 80$, or when $x \geq 80$. Hence,

$$\{(x, y)|x < 80 \rightarrow y > 80\}$$
$$= \{(x, y)|x < 80 \quad \text{and} \quad y > 80, \text{ or } x \geq 80\}$$
$$= \{(75, 90), (85, 90), (90, 85), (80, 90), (100, 95), (80, 95)\}$$

(2) $\{(x, y)|x \geq 75 \rightarrow y > 90\}$

$= \{(x, y)|x \geq 75 \quad \text{and} \quad y > 90, \text{ or } x < 75\}$

$= \{(100, 95), (80, 95), (65, 75), (65, 80), (70, 80)\}.$

(3)   Here we look for those students whose scores on both tests were below 80, or both not below 80, giving

$\{(x, y)|x < 80 \Leftrightarrow y < 80\} =$

$\{(75, 60), (75, 75), (85, 90), (90, 85), (100, 95)\}$

EXERCISES

In Exercises 1–12, $P = \{x|p(x)\}$ and $Q = \{x|q(x)\}$. Which of the relations are true if $p(x) \rightarrow q(x)$?

1. $P \cap Q = \emptyset$ 2. $P = Q$
3. $\overline{P} \subset \overline{Q}$ 4. $\overline{Q} \subset \overline{P}$
5. $P \cap \overline{Q} = \emptyset$ 6. $\overline{P} \cap \overline{Q} = \overline{Q}$

In the exercises below, decide if $p(x) \rightarrow q(x)$, $q(x) \rightarrow p(x)$, $p(x) \Leftrightarrow q(x)$, or if $p(x)$ and $q(x)$ are unrelated in any of these ways.

7. $P \cap Q = \emptyset$ 8. $P \cup Q = P$
9. $P \cap Q = Q$ 10. $P \subset \overline{Q}$
11. $\overline{P} \subset \overline{Q}$ 12. $P \cap \overline{Q} = \emptyset$

The universal set in Exercises 13–18 is the set of real numbers. Verify the given statements, using truth sets.

13. $x^2 = x \Leftrightarrow x = 0 \text{ or } x = 1$
14. $x^2 - 3x + 2 = 0 \rightarrow x^3 - 3x^2 + 2x = 0$
15. $x < -3 \rightarrow x < -2$
16. $x < 1 \rightarrow x \leq 1$
17. $x > 2 \rightarrow x > 1$
18. $x = -x \Leftrightarrow x = 0$

In Table 1.4 each year (y) has *exactly one* pair (w, ℓ) associated with it, and each club (c) has *at least* one such pair associated with it. Find the members of the sets in Exercises 19–24.

19. $\{y|w > 100 \Leftrightarrow \ell > 50\}$
20. $\{y|w < 95 \Leftrightarrow \ell < 60\}$
21. $\{y|w > 100 \Leftrightarrow \ell > 60\}$

TABLE 1.4 NATIONAL LEAGUE PENNANT WINNERS, 1940–1969

Year (y)	Club (c)	Won (w)	Lost (ℓ)	Year (y)	Club (c)	Won (w)	Lost (ℓ)
1940	Cincinnati	100	53	1955	Brooklyn	98	55
1941	Brooklyn	100	54	1956	Brooklyn	93	61
1942	St. Louis	106	48	1957	Milwaukee	95	59
1943	St. Louis	105	49	1958	Milwaukee	92	62
1944	St. Louis	105	49	1959	Los Angeles	88	68
1945	Chicago	98	56	1960	Pittsburgh	95	59
1946	St. Louis	98	58	1961	Cincinnati	93	61
1947	Brooklyn	94	60	1962	San Francisco	103	62
1948	Boston	91	62	1963	Los Angeles	99	63
1949	Brooklyn	97	57	1964	St. Louis	93	69
1950	Philadelphia	91	63	1965	Los Angeles	97	65
1951	New York	98	59	1966	Los Angeles	95	67
1952	Brooklyn	96	57	1967	St. Louis	107	60
1953	Brooklyn	105	49	1968	St. Louis	97	65
1954	New York	97	57	1969	New York	100	62

22. $\{c | w < 100 \rightarrow y > 1964\}$
23. $\{c | w > 93 \rightarrow y < 1945\}$
24. $\{c | 1950 < y < 1960 \rightarrow \ell < 59\}$

Express the sets in Exercises 25–31 in as simple a form as you can.

25. $(\overline{A} \cap B) \cup (A \cap \overline{B})$
26. $(A \cap B) \cup (A \cap \overline{B})$
27. $(A \cup B) \cap (A \cup \overline{B})$
28. $[A \cup (\overline{B} \cap C)] \cap [A \cup (B \cap C)]$
29. $(A - B) \cap (A \cup \overline{B})$
30. $(A - B) \cap (\overline{A} \cup B)$
31. $\overline{(A - B)}$

32. A labor union has adopted the following policies:

   (a) The financial committee is to be elected from among the union leadership.
   (b) No member of the financial committee can serve on the appropriations committee.

   Let L stand for the union leadership, F for the finance committee, and A for the appropriations committee. Express statements (a) and (b) in set notation.

33. A coin is tossed three times. Consider the statements

$p(x)$: The outcome x is at least one Head
$q(x)$: The outcome x is two Heads and one Tail
$r(x)$: The outcome x is two or more Heads.

Which of the following implications are correct?

(a) $p(x) \rightarrow q(x)$
(b) $q(x) \rightarrow p(x)$
(c) $r(x) \rightarrow p(x)$

34. This exercise continues Exercise 33. Find the following truth sets:

(a) $P = \{x|p(x)\}$ (b) $Q = \{x|q(x)\}$
(c) $R = \{x|r(x)\}$ (d) $P \cap Q$
(e) $Q \cap R$

2
BASIC SYMBOLIC LOGIC

2

In deciding if a given statement is true or false we have to go through some form of reasoning. At times, when the statement under consideration is very complicated, we may be unable to reach a decision because of the haphazardness of our reasoning, or logic. At other times we arrive at the wrong answer. When this happens we say that our *logic* was at fault. What this logic refers to is the process of correct inference. Our main goal in this chapter is to develop such a process. This will be achieved by converting verbal statements into symbolic ones and applying our knowledge of sets.

2.1 STATEMENTS AND TRUTH SETS

The arguments in this chapter are based on two simple assumptions:

(1) Each occurring statement is either true *or* false.
(2) No statement is both true *and* false.

This means that we take it for granted that any question of the form "Is statement $p(x)$ true?" can always be answered yes or no. To further illustrate the relation between statements and their truth sets (see Section 1.3), let us study an example in detail.

Example 1

Table 2.1 lists the status of three grounds for divorce in eight states. Referring to this table, consider the following statements in which x stands for a state:

$p(x)$ means "pregnancy at marriage is grounds for divorce in x"
$q(x)$ means "nonsupport is grounds for divorce in x"
$r(x)$ means "mental cruelty is grounds for divorce in x"

TABLE 2.1   GROUNDS FOR DIVORCE IN SELECTED
STATES (1968)

State	Pregnant at marriage	Nonsupport	Mental cruelty
Florida	No	No	Yes
Louisiana	No	No	No
Massachusetts	No	Yes	No
Michigan	No	Yes	Yes
New Mexico	Yes	Yes	Yes
Pennsylvania	Yes	No	No
Tennessee	Yes	Yes	Yes
Vermont	No	Yes	No
Wisconsin	No	Yes	Yes

By Section 1.3 the set

$$P = \{x|p(x)\}$$

is the set of all states x for which $p(x)$ is true. We thus take for P the
set of states x having "yes" in the "Pregnant at marriage" column
(Table 2.1):

$$P = \{\text{New Mexico, Pennsylvania, Tennessee}\}$$

This is the *truth set* of the statement $p(x)$. Similarly, if

$$Q = \{x|q(x)\} \quad \text{and} \quad R = \{x|r(x)\}$$

then we find from Table 2.1 that

$$Q = \{\text{Massachusetts, Michigan, New Mexico, Tennessee,}$$
$$\text{Vermont, Wisconsin}\}$$
$$R = \{\text{Florida, Michigan, New Mexico, Tennessee, Wisconsin}\}$$

The relations between P, Q, and R are described graphically in Figure
2.1, where the universal set $\mathfrak{U}$ consists of all nine states in Table 2.1.
Note that each of the statements $p(x)$, $q(x)$, and $r(x)$ is false when
$x = $ Louisiana.

Consider the statement, "Pregnancy at marriage *and* non-

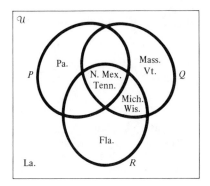

FIGURE 2.1

support are grounds for divorce in x." In symbolic form this statement can be expressed as

$$p(x) \quad \text{and} \quad q(x)$$

We see that for this statement to be true, $p(x)$ must be true and $q(x)$ must be true. This is the case when $x = $ New Mexico or $x = $ Tennessee. Since

$$P \cap Q = \{\text{New Mexico, Tennessee}\}$$

(see Figure 2.1), we find that the statement "$p(x)$ and $q(x)$" is true for $x \in P \cap Q$ and false for $x \notin P \cap Q$. From this we conclude that

$$P \cap Q \text{ is the } \textit{truth set} \text{ of the statement } "p(x) \text{ and } q(x)"$$

Example 2
Write the following statement in symbolic form and find its truth set: "Pregnancy at marriage *or* mental cruelty are grounds for divorce in x."

SOLUTION
This statement can be expressed in the form

$$p(x) \quad \text{or} \quad r(x)$$

It is true when the state x has a "yes" in the "pregnant at marriage" column or a "yes" in the "nonsupport" column. From the point of view of sets, this statement is true when $x \in P$ or $x \in R$, which is the same as saying that the statement is true when $x \in P \cup R$. We see at once that "$p(x)$ or $r(x)$" is false for $x \notin P \cup R$, and hence the truth set of this statement is

$$P \cup R = \{\text{Florida, Michigan, New Mexico, Pennsylvania,}$$
$$\text{Tennessee, Wisconsin}\}$$

Example 3

For what states x is the following statement true (according to Table 2.1): "The only grounds for divorce in x are pregnancy at marriage"?

SOLUTION

The states which qualify are those having "yes" in the "pregnant at marriage" column and "no" in each of the other two columns. It is seen that the statement is true if and only if x = Pennsylvania, and so its truth set is

$$\{\text{Pennsylvania}\} = P - (Q \cup R)$$

Example 4

In what states are the grounds for divorce "pregnancy at marriage and nonsupport and mental cruelty"?

SOLUTION

The states we are seeking are those states for which the three statements $p(x)$, $q(x)$, and $r(x)$ are true. These are precisely the states having "yes" in every column; expressed in the language of sets, this says that $x \in P \cap Q \cap R$. From Figure 2.1 we find that

$$P \cap Q \cap R = \{\text{New Mexico, Tennessee}\}$$

Example 5

For what states x is the following statement true? "Nonsupport and mental cruelty are *not* grounds for divorce."

SOLUTION

Using the language of sets, we can say that this statement is true if and only if $x \notin Q$ and $x \notin R$. This leaves the two possibilities x = Pennsylvania or x = Louisiana. Hence, the truth set is

$$\{\text{Pennsylvania, Louisiana}\}$$

From Figure 2.1 we see that this set is precisely

$$\mathfrak{U} - (Q \cup R) = \overline{Q \cup R}$$

Example 6

Give a verbal description of the set P.

SOLUTION

P is the set of states in which pregnancy at marriage is grounds for divorce.

Example 7

Give a verbal description of the statement "not $r(x)$."

SOLUTION

By the statement "not $r(x)$" we mean "$r(x)$ is false." Hence, it says: "Mental cruelty is not grounds for divorce in x."

EXERCISES

Referring to Table 2.1, list the states for which each of the statements in Exercises 1–7 is true.

1. Grounds for divorce are pregnancy at marriage and mental cruelty.
2. Grounds for divorce are nonsupport or mental cruelty.
3. Nonsupport is not grounds for divorce.
4. Pregnancy at marriage and nonsupport and mental cruelty are not grounds for divorce.
5. Pregnancy at marriage and nonsupport are not grounds for divorce.
6. Nonsupport is the only grounds for divorce.
7. Give the truth sets of the statements in Exercises 1–6 in terms P, Q, and R, and set operations.

Referring to Example 1: Give a verbal equivalent of the statements in Exercises 8–12.

8. $q(x)$ or $r(x)$
9. $q(x)$ and not $r(x)$
10. Not $p(x)$
11. Not $p(x)$ and not $q(x)$
12. $p(x)$ and $q(x)$ or $p(x)$ and $r(x)$

Give a verbal description of the sets in Exercises 13–17.

13. R
14. $P \cup R$

15. $\overline{P}$
16. $\overline{P \cup Q}$
17. $Q - R$

18. $\mathfrak{U} - \{\text{Louisiana}\}$

19. A die is rolled. Consider the statements

$p(x)$: The outcome x is an even number.
$q(x)$: The outcome x is an odd number.
$r(x)$: The outcome x is a number less than 4.

Find the following truth sets:

(a) $P = \{x | p(x)\}$
(b) $Q = \{x | q(x)\}$
(c) $R = \{x | r(x)\}$
(d) $P \cap Q$
(e) $P \cap R$
(f) $\overline{P} \cup R$

20. This exercise continues Exercise 19. Give a verbal interpretation of the following sets:

(a) $\overline{P}$
(b) $R \cap \overline{P}$
(c) $\overline{P} \cap \overline{R}$
(d) $\overline{P \cap R}$

2.2   CONNECTIVES

Recall from Section 1.3 how the connectives *and*, *or*, and *not* are associated with sets:

$$P \cap Q = \{x | x \in P \quad \text{and} \quad x \in Q\}$$
$$P \cup Q = \{x | x \in P \quad \text{or} \quad x \in Q\}$$
$$\overline{P} = \{x | x \text{ is not in } P\}$$

To study these connectives from the point of view of logic, we introduce new symbols which are compared with set operations in Table 2.2. A statement $p \wedge q$ is called the *conjunction* of p and q; $p \vee q$ is called the *disjunction* of p and q; $\sim p$ is called the *negation* of p. The usage of these new symbols is explained in the examples below, where we designate statements by $p, q, r, \ldots$ rather than $p(x), q(x), r(x), \ldots$

TABLE 2.2 COMMON CONNECTIVES

Connective	Logic notation		Set notation	
and	$\wedge$	conjunction	$\cap$	intersection
or	$\vee$	disjunction	$\cup$	union
not	$\sim$	negation	$-$	complementation

Example 1
Let p stand for "two points determine a line," q for "a line is determined by any two points on it." Then

$$p \wedge q \quad \text{read} \quad \text{"}p \text{ and } q\text{"}$$

stands for "two points determine a line and a line is determined by any two points on it."

Example 2
Consider the statement

"I still love art and life very much." (Van Gogh)

Letting p stand for "I still love art very much," q for "I still love life very much," this statement can be expressed symbolically as $p \wedge q$. Observe that we are interested only in the meaning of a statement and not in its grammatical form.

Example 3
Let p stand for "to be." Then

$$\sim p \quad \text{read} \quad \text{"not } p\text{"}$$

stands for "not to be." Hence,

$$p \vee \sim p \text{ stands for "to be or not to be"}$$

Example 4
Let p stand for "the team won the first game," q for "the team won the second game." Give a verbal interpretation of the statements:

(1) $p \wedge q$
(2) $\sim p \wedge q$
(3) $\sim p \wedge \sim q$

SOLUTION

(1) "The team won the first and the second game."
(2) "The team lost the first and won the second game."
(3) "The team lost the first and the second game."

Statements of the form $p \vee q$, $\sim p$, and $\sim p \wedge q$, are examples of
compound statements; p and q are then *simple statements*. We are using
these terms in a relative sense, and no precise definitions are needed.
To find the truth sets of the basic statements $p \wedge q$, $p \vee q$, and $\sim p$,
consult Figure 2.2, in which P is the truth set of p and Q is the truth
set of q.

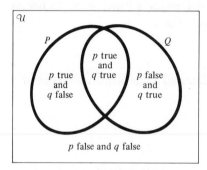

FIGURE 2.2   *Truth sets*

With the observation that

      "p is false" means the same as "not p is true"

we obtain the truth sets listed in Table 2.3.

TABLE 2.3   TRUTH SETS OF
SELECTED STATEMENTS

Statement	Truth set
p	P
q	Q
$p \wedge q$	$P \cap Q$
$p \vee q$	$P \cup Q$
$\sim p$	$\bar{P}$
$p \wedge \sim q$	$P \cap \bar{Q}$
$\sim p \wedge q$	$\bar{P} \cap Q$
$\sim p \wedge \sim q$	$\bar{P} \cap \bar{Q}$

Example 5
Find the truth set of $p \vee \sim q$.

SOLUTION
This is done with cross-hatching in a Venn diagram. If the truth set of p is P and that of q is Q, then the truth set of $\sim q$ is $\bar{Q}$. The truth set of $p \vee \sim q$ is seen to be the *total shaded area* in Figure 2.3. Hence,

the truth set of $p \vee \sim q$ is $P \cup \bar{Q}$

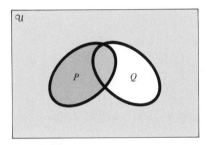

FIGURE 2.3   *The truth set of* p $\vee \sim$q

Example 6
Find the truth sets of $p \wedge \sim p$ and $p \vee \sim p$.

SOLUTION
If the truth set of p is P, then

the truth set of $p \wedge \sim p$ is $P \cap \bar{P} = \emptyset$
the truth set of $p \vee \sim p$ is $P \cup \bar{P} = \mathcal{U}$

Thus, $p \wedge \sim p$ is never true, whereas $p \vee \sim p$ is always true.

Terminology

A statement which is always false is said to be a *self-contradiction*.
A statement which is always true is said to be a *tautology*.
$p \wedge \sim p$ is an example of a self-contradiction and $p \vee \sim p$ is an example of a tautology.

EXERCISES

1. Let p stand for "the exercise is long," q for "the exercise is easy." Give a verbal equivalent of each of the following statements.

    (a) $p \wedge q$
    (b) $\sim p$

    (c) $\sim p \wedge q$
    (d) $\sim p \wedge \sim q$
    (e) $\sim (p \wedge q)$
    (f) $\sim (p \wedge \sim q)$

2.   Let p stand for "production expenses are high," q for "profits are high." Give a verbal equivalent of each of the following statements.

    (a) $p \vee q$
    (b) $p \wedge \sim q$
    (c) $p \wedge (q \vee \sim q)$
    (d) $\sim p \wedge \sim q$
    (e) $\sim (\sim p \vee \sim q)$
    (f) $(p \wedge \sim q) \vee (\sim p \wedge q)$

3.   Let p stand for "look," q for "listen." Write each of the following statements in symbolic form.

    (a)   Look and listen.
    (b)   Do not look or listen.
    (c)   Do not look and do not listen.
    (d)   Look but do not listen.
    (e)   Listen or do not, but look.

Write each of the following compound statements in symbolic form and give their simple components.

4.   The inequality $a < b$ holds or the inequality $b \leq a$ holds.
5.   An integer is zero, even, or odd.
6.   The statement $a = -a$ is false, or $a = 0$ is true.
7.   The answer to question 1 was "yes," but the answer to question 2 was "no."
8.   The time is 10 or 11 o'clock and it is late.

Exercises 9–13 refer to Table 2.4. Find the truth set of each statement. x stands for a planet.

9. x has escape velocity not exceeding 4 miles per second.
10. The distance of x to the sun exceeds five astronomical units, and its period of revolution around the sun is less than 20 years.
11. x has escape velocity in excess of 10 miles per second, and its distance from the sun is less than two astronomical units.

12. The period of revolution of x around the sun is at least a year, its escape velocity is between 6 and 30 miles per second, and its distance from the sun exceeds ten astronomical units.
13. The number giving the escape velocity of x is less than the number giving its distance from the sun.

TABLE 2.4 SELECTED PLANETARY DATA

Planet (x)	Distance from the sun in astronomical units[1]	Period of revolution around the sun	Escape velocity in miles per second[2]
Mercury	0.387	87.96 days	2.60
Venus	0.723	224.70 days	6.38
Earth	1.000	365.25 days	6.94
Mars	1.524	1.88 years	3.10
Jupiter	5.203	11.86 years	37.82
Saturn	9.540	29.45 years	22.94
Uranus	19.180	84.01 years	13.64
Neptune	30.070	164.79 years	15.50
Pluto	39.440	248.40 years	6.20

[1] An **astronomical unit** is approximately 93 million miles and it gives the mean distance between earth and sun.
[2] **Escape velocity** is the velocity necessary to overcome the gravitational pull to escape into space.

Let p, q, and r have truth sets P, Q, and R, respectively. Find the truth sets of the statements in Exercises 14–24. You are reminded of the list of properties of unions and intersections in Section 1.1 and the list of properties of complements and relative complements in Section 1.2.

14. $p \wedge q \wedge r$
15. $p \wedge q \wedge \sim r$
16. $p \wedge (\sim q \wedge \sim r)$
17. $p \wedge (\sim q \vee \sim r)$
18. $p \vee (q \wedge r)$
19. $p \wedge (q \vee r)$
20. $\sim p \vee \sim q \vee \sim r$
21. $\sim (p \vee q)$
22. $\sim (\sim p \vee \sim q)$
23. $\sim [p \vee (q \wedge r)]$
24. $(p \wedge \sim q) \vee (\sim p \wedge q)$

25. A coin is tossed four times. Let p stand for "the outcome was two

Heads," q for "the outcome was three Heads." Give a verbal interpretation of the following statements:

(a) $p \lor q$
(b) $\sim q$
(c) $\sim p \land \sim q$
(d) $\sim (p \lor q)$

2.3 TRUTH TABLES

In Section 2.1 we stipulated that any occurring statement p is either true or false, and these two possible answers are said to be the *truth values* of p. In a compound statement the truth values T = True and F = False can be assigned to simple component statements in various ways. Each such assignment gives a truth value for the compound statement. The set of all distinct assignments of truth values to simple statements gives the *logical possibilities* of the compound statement composed of them.

Example 1

A statement composed of p and q has four possible assignments of truth values to p and q. These are listed in Table 2.5.

TABLE 2.5   THE LOGICAL POSSIBILITIES OF A STATEMENT COMPOSED OF p AND q

p	q
T	T
T	F
F	T
F	F

Example 2

A statement composed of p, q, and r has eight possible assignments of truth values. These logical possibilities are listed in Table 2.6.

We can now give a more satisfactory definition of the terms *tautology* and *self-contradiction* introduced in Section 2.2. Namely, a *tautology* is a statement that is *true* for all logical possibilities; such a statement is also said to be *logically true*. A *self-contradiction* is a statement which is *false* for all logical possibilities; such a statement is also called *logically false*.

TABLE 2.6   THE LOGICAL POSSIBILITIES OF
A STATEMENT COMPOSED OF p, q, AND r

p	q	r
T	T	T
T	F	T
F	T	T
F	F	T
T	T	F
T	F	F
F	T	F
F	F	F

The truth sets discussed in the preceding two sections enable us to decide how the truth values of a compound statement involving the connectives $\wedge$, $\vee$, and $\sim$ depend on the truth values of its simple component statements. From Figure 2.4 we easily get the truth values of the statements $p \wedge q$ and $p \vee q$ for all logical possibilities. These are listed in Table 2.7, called a *truth table*. From Figure 2.4 we also ob-

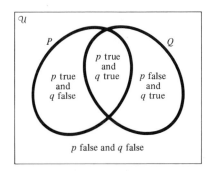

FIGURE 2.4   *Truth sets*

TABLE 2.7   THE TRUTH TABLE OF $p \wedge q$ AND $p \vee q$

p	q	$p \wedge q$	$p \vee q$
T	T	T	T
T	F	F	T
F	T	F	T
F	F	F	F

tain the truth table of $\sim p$ (see Table 2.8). With these basic truth tables we can analyze more complex statements, using a step-by-step procedure as explained in the examples below.

TABLE 2.8   THE TRUTH TABLE OF $\sim p$

p	$\sim p$
T	F
F	T

Example 3

Construct the truth table of the statement $p \vee \sim q$.

SOLUTION

Step 0   Construct Table 2.9(a), listing the logical possibilities of p and q.

Step 1   Write the truth values of p and $\sim q$ in the right side of the table [see Table 2.9(b)].

TABLE 2.9(a)

p	q	p	$\vee$	$\sim q$
T	T			
T	F			
F	T			
F	F			
↑	↑			
⓪	⓪			

TABLE 2.9(b)

p	q	p	$\vee$	$\sim q$
T	T	T		F
T	F	T		T
F	T	F		F
F	F	F		T
		↑		↑
		①		①

Step 2 Use the columns obtained in Step 1 to read off the truth values from Table 2.7. The truth values of $p \vee \sim q$ are given in the center column of Table 2.9(c).

TABLE 2.9(c)   THE TRUTH TABLE OF $p \vee \sim q$

p	q	p	$\vee$	$\sim q$
T	T	T	T	F
T	F	T	T	T
F	T	F	F	F
F	F	F	T	T

Once we are familiar with the procedure, we can carry out all the steps in a single table. This is done below.

Example 4
Construct the truth table of the statement $(p \vee q) \wedge (\sim p \vee q)$.

SOLUTION
As with algebraic manipulations we first evaluate each statement inside parentheses.

Step 1 Enter in Table 2.10 the truth values of p, q, and $\sim p$.
Step 2 Use Step 1 to evaluate $(p \vee q)$ and $(\sim p \vee q)$.

TABLE 2.10   THE TRUTH TABLE OF $(p \vee q) \wedge (\sim p \wedge q)$

p	q	$(p$	$\vee$	$q)$	$\wedge$	$(\sim p$	$\vee$	$q)$
T	T	T	T	T	T	F	T	T
T	F	T	T	F	F	F	F	F
F	T	F	T	T	T	T	T	T
F	F	F	F	F	F	T	T	F

Step 3   Use Step 2 to find the truth values of the complete statement. Observe that this statement is true when q is true, and false when q is false. Indeed, using a Venn diagram you will easily see that the truth set of this statement is the truth set of q.

Example 5
Construct the truth table of the statement

$$[(p \land q) \lor (p \land r)] \lor \sim(p \lor r)$$

SOLUTION

Step 1   Enter in Table 2.11 the truth values of p, q, and r which correspond to the listing of the logical possibilities in the left side.

Step 2   Find the truth values of $(p \land q)$, $(p \land r)$, and $(p \lor r)$.

Step 3   Find the truth values of $[(p \land q) \lor (p \land r)]$ and $\sim(p \lor r)$.

Step 4   Find the truth values of the complete statement.

TABLE 2.11   THE TRUTH TABLE OF
$[(p \land q) \lor (p \land r)] \lor \sim(p \lor r)$

p	q	r	[(p	∧	q)	∨	(p	∧	r)]	∨	~	(p	∨	r)
T	T	T	T	T	T	T	T	T	T	T	F	T	T	T
T	F	T	T	F	F	T	T	T	T	T	F	T	T	T
F	T	T	F	F	T	F	F	F	T	F	F	F	T	T
F	F	T	F	F	F	F	F	F	T	F	F	F	T	T
T	T	F	T	T	T	T	T	F	F	T	F	T	T	F
T	F	F	T	F	F	F	T	F	F	F	F	T	T	F
F	T	F	F	F	T	F	F	F	F	T	T	F	F	F
F	F	F	F	F	F	F	F	F	F	F	T	F	F	F

Example 6
Construct the truth table of the statement

$$[p \lor (\sim p \land q)] \lor \sim q$$

SOLUTION

Following the procedure described in the preceding problems you should be able to verify correctness of Table 2.12 without difficulty. Study carefully the way in which the steps were carried out. Observe that this statement is logically true.

TABLE 2.12   THE TRUTH TABLE OF $[p \vee (\sim p \wedge q)] \vee \sim q$

---

p	q	$[p$	$\vee$	$(\sim p$	$\wedge$	$q)]$	$\vee$	$\sim q$
T	T	T	T	F	F	T	T	F
T	F	T	T	F	F	F	T	T
F	T	F	T	T	T	T	T	F
F	F	F	F	T	F	F	T	T

EXERCISES

Construct truth tables for the compound statements in Exercises 1–12.

1. $p \wedge \sim q$
2. $\sim(p \wedge \sim q)$
3. $\sim(p \vee q)$
4. $(p \wedge \sim q) \vee (\sim p \wedge q)$
5. $\sim(p \vee q) \wedge (\sim p \vee q)$
6. $p \wedge [(q \wedge r) \vee \sim p]$
7. $(p \vee q) \wedge \sim(p \wedge r)$
8. $[\sim p \wedge (p \vee q)] \wedge r$
9. $(p \wedge q) \vee (p \wedge r)$
10. $\sim(p \wedge q) \vee \sim(p \wedge r)$
11. $\sim(\sim p \vee \sim q)$
12. $\sim(\sim p \wedge \sim q)$

13. Joan said, "My answers to questions A and B were wrong, or my answers to questions B and C were wrong." What is the largest score Joan could get if each question counted 1 point?
14. Jack said, "I did not answer questions A or B, and I did not answer questions A or C." What questions did Jack answer?

Decide if the following statements (Exercises 15 and 16) are true or false:

15. It is not true that parallel or coincident lines do not have exactly one point in common.
16. It is not true that 5 is not odd and 5 is a prime number.
17. List the logical possibilities of a statement composed of p, q, r, and t.
18. List the logical possibilities of a statement composed of p, q, r, s, and t in the following cases:

    (a) s and t are true
    (b) s and t are true and p is false
    (c) s is false or t is false and p, q, and r are false

Two statements are equivalent if they have the same truth set or truth table. Show that the statements in Exercise 19–25 are equivalent.

19. $\sim(p \wedge q)$ and $\sim p \vee \sim q$
20. $\sim(p \vee q)$ and $\sim p \wedge \sim q$
21. $\sim(\sim p)$ and p
22. $\sim(\sim p \vee q)$ and $p \wedge \sim q$
23. $\sim(p \wedge \sim q)$ and $\sim p \vee q$
24. $(p \wedge q) \vee r$ and $(p \vee r) \wedge (q \vee r)$
25. $(p \vee q) \wedge r$ and $(p \wedge r) \vee (q \wedge r)$

2.4 THE CONDITIONAL AND BICONDITIONAL CONNECTIVES

In Section 1.4 we introduced the notation $p \Leftrightarrow q$ for statements having the same truth set (equivalent statements), and the notation $p \rightarrow q$ for the case when the truth set of p is a subset of the truth set of q (p implies q). In logic we call the symbol $\Leftrightarrow$ *biconditional connective* and the symbol $\rightarrow$ *conditional connective*. The two basic facts to recall from Section 1.4 are these:

$p \Leftrightarrow q$ stands for "p if and only if q"
$p \rightarrow q$ stands for "if p then q"

Example 1

Let p stand for "$(x - 1)(x - 2) = 0$," q for "$x = 1$ or $x = 2$." Then $p \Leftrightarrow q$ stands for the statement

"$(x - 1)(x - 2) = 0$ if and only if $x = 1$ or $x = 2$"

Alternative ways for saying this are

1. $(x - 1)(x - 2) = 0$ if $x = 1$ or $x = 2$

and conversely,

$x = 1$ or $x = 2$ if $(x - 1)(x - 2) = 0$

2.   if $(x - 1)(x - 2) = 0$, then $x = 1$ or $x = 2$

and

if $x = 1$ or $x = 2$, then $(x - 1)(x - 2) = 0$

Example 2

Let p stand for "two lines intersect," q for "two lines are parallel." Then $p \rightarrow \sim q$ stands for

"if two lines intersect then they are not parallel"

According to Section 1.4, the relation $p \Leftrightarrow q$ signifies that p and q have the same truth sets. This means that either p and q are both true, or p and q are both false. From this we see that the truth table (Table 2.13) of $p \Leftrightarrow q$ is as follows:

TABLE 2.13   THE TRUTH TABLE OF THE STATEMENT $p \Leftrightarrow q$

p	q	$p \Leftrightarrow q$
T	T	T
T	F	F
F	T	F
F	F	T

To obtain the truth table of $p \rightarrow q$, we use the fact that

$p \Leftrightarrow q$ is the same as $(p \rightarrow q) \wedge (q \rightarrow p)$

(see Section 1.4). We can thus write Table 2.13 in the following form (Table 2.14):

TABLE 2.14

p	q	$(p \rightarrow q) \wedge (q \rightarrow p)$
T	T	T
T	F	F
F	T	F
F	F	T

Some of the missing truth values can be deduced from the observation that $r \wedge s$ is true if and only if r is true and s is true. This leads us to Table 2.14(a), part of which is shaded in, since we are interested only in the truth values of $p \rightarrow q$. Since the truth set of p is a subset of the truth set of q when $p \rightarrow q$, it follows that $p \rightarrow q$ is false when p is true and q is false. This leads us to Table 2.14(b).

We now arrive at an impasse. Our fund of knowledge of logic and sets is exhausted, but we are left with an incomplete truth table. This

TABLE 2.14(a)

p	q	$(p \rightarrow q)$	$\wedge$	$(q \rightarrow p)$
T	T	T	T	T
T	F		F	
F	T		F	
F	F	T	T	T

TABLE 2.14(b)

p	q	$p \rightarrow q$
T	T	T
T	F	F
F	T	
F	F	T

state of affairs is very unsatisfactory because we work with the assumption that a truth value can be assigned to *any* statement that comes up. To get out of this dilemma, we *arbitrarily* assign a truth value to the statement $p \rightarrow q$ when the assumption p is false and the conclusion q is true. The truth value we assign is T, and this permits us to construct Table 2.15.

TABLE 2.15   THE TRUTH TABLE
OF THE STATEMENT $p \rightarrow q$

p	q	$p \rightarrow q$
T	T	T
T	F	F
F	T	T
F	F	T

Warning

A statement such as

"if $1 = 0$ then $2^2 = 4$"

is true because we *defined* it to be true and *not* because of any logical reasoning.

Truth tables involving the connectives introduced in this section are constructed as before. We therefore omit much of the discussion in the examples below.

Example 3
Construct the truth table of the statement $p \Leftrightarrow (p \rightarrow \sim q)$.

SOLUTION
The construction of Table 2.16 is easily understood by following the indicated steps. As usual, we first assign truth values to statements inside parentheses.

Example 4
Construct the truth table of the statement $(p \Leftrightarrow q) \rightarrow \sim(p \wedge q)$; (Table 2.17).

TABLE 2.16   THE TRUTH TABLE OF $p \Leftrightarrow (p \rightarrow \sim p)$

p	q	p	$\Leftrightarrow$	(q	$\rightarrow$	$\sim p$)
T	T	T	F	T	F	F
T	F	T	T	F	T	F
F	T	F	F	T	T	T
F	F	F	F	F	T	T

TABLE 2.17   THE TRUTH TABLE OF $(p \Leftrightarrow q) \rightarrow \sim(p \wedge q)$

p	q	(p	$\Leftrightarrow$	q)	$\rightarrow$	$\sim$	(p	$\wedge$	q)
T	T	T	T	T	F	F	T	T	T
T	F	T	F	F	T	T	T	F	F
F	T	F	F	T	T	T	F	F	T
F	F	F	T	F	T	T	F	F	F

SOLUTION

Again follow the steps at the bottom of the table (2.17).

Example 5

Show that the statements $p \rightarrow q$ and $\sim p \vee q$ are equivalent.

SOLUTION

We recall from Section 1.4 that two statements are equivalent if and only if they have the same truth set. But statements having the same truth set must be true together or false together, and hence they must have the same truth table. This is shown in Table 2.18 to be indeed the case (in this connection see Exercises 19–25 in Section 2.3).

TABLE 2.18   THE TRUTH TABLES OF THE
EQUIVALENT STATEMENTS $p \to q$ AND $\sim p \vee q$

p	q	$p \to q$	$\sim p$	$\vee$	q
T	T	T	F	T	T
T	F	F	F	F	F
F	T	T	T	T	T
F	F	T	T	T	F

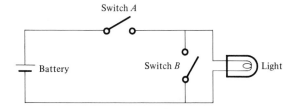

Example 6

Consider the wiring diagram in Figure 2.5. For the light to be on, switch A must be closed and switch B must be open. When is the following statement true?

"The light is on if and only if switch A is closed and switch B is open."

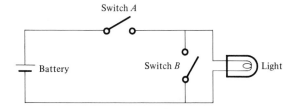

Switch A

Battery   *Switch B*   Light

FIGURE 2.5   *Wiring diagram*

SOLUTION

Consider the following simple statements:

ℓ stands for "the light is on."
a stands for "switch A is closed."
b stands for "switch B is closed."

The statement in the problem can be written symbolically as

$$\ell \Leftrightarrow (a \wedge \sim b)$$

TABLE 2.19 THE TRUTH TABLE
OF THE STATEMENT $\ell \Leftrightarrow (a \land \sim b)$

ℓ	a	b		ℓ	$\Leftrightarrow$	$(a$	$\land$	$\sim b)$
T	T	T		T	F	T	F	F
T	F	T		T	F	F	F	F
F	T	T		F	T	T	F	F
F	F	T		F	T	F	F	F
T	T	F		T	T	T	T	T
T	F	F		T	F	F	F	T
F	T	F		F	F	T	T	T
F	F	F		F	T	F	F	T

To find when this statement is true, we construct its truth table (Table 2.19).

In Table 2.19 we shaded in the logical possibilities for which the statement is false. The logical possibilities for which the statement is true are listed verbally below.

Light on	Switch A closed	Switch B closed
False	True	True
False	False	True
True	True	False
False	False	False

EXERCISES

Construct truth tables for the statements in Exercises 1–12.

1. $\sim p \rightarrow q$
2. $p \rightarrow \sim q$
3. $\sim p \rightarrow \sim q$
4. $(p \rightarrow \sim p) \rightarrow \sim p$

   5. $(p \wedge q) \rightarrow q$

   6. $(\sim p \wedge \sim q) \rightarrow \sim (p \wedge q)$

   7. $(\sim p \wedge \sim q) \Leftrightarrow (p \vee q)$

   8. $(p \Leftrightarrow q) \rightarrow [(p \vee q) \vee (\sim p \wedge \sim q)]$

   9. $(p \Leftrightarrow q) \rightarrow (p \Leftrightarrow r)$

 10. $(p \rightarrow q) \rightarrow (q \rightarrow r)$

 11. $(p \rightarrow q) \rightarrow (p \rightarrow r)$

 12. $(p \vee q) \Leftrightarrow (p \vee r)$

13.   The statement $\sim q \rightarrow \sim p$ is said to be the *contrapositive* of $p \rightarrow q$. Show that $p \rightarrow q$ and $\sim q \rightarrow \sim p$ are equivalent.

14.   Which of the following statements is equivalent to $p \Leftrightarrow q$?

$$\sim p \Leftrightarrow \sim q \quad (p \rightarrow \sim q) \wedge (\sim p \rightarrow q) \quad (\sim p \wedge q) \wedge (p \wedge \sim q)$$

15.   Which of the following statements is equivalent to $q \rightarrow p$?

$$\sim p \rightarrow \sim q \quad q \vee \sim p \quad \sim q \vee p \quad \sim (\sim p \wedge q)$$

16.   Which of the following statements are logically true (tautologies)?

$$p \Leftrightarrow (p \vee q) \quad p \rightarrow (p \vee q) \quad (\sim p \wedge \sim q) \rightarrow \sim p$$

17.   Consider the wiring diagram in Figure 2.6. For light A to be on, switch A must be closed, and for light B to be on, both switches A *and* B must be closed. When is the following statement true?

"Lights A and B are on if and only if switches A and B are closed."

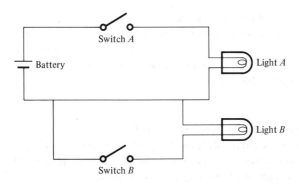

FIGURE 2.6

18.   Consider the wiring diagram in Figure 2.7. For light A to be on,

switch A must be closed, and for light B to be on, switch B must be closed. When is the following statement true?

"If switch B is closed, then lights A and B are on."

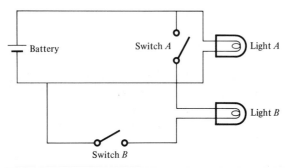

FIGURE 2.7

Using truth sets or truth tables, show that the statements in each exercise below are equivalent.

19. $p \rightarrow q$ and $\sim p \lor q$
20. $p \Leftrightarrow q$ and $(p \land q) \lor (\sim p \land \sim q)$
21. $\sim(p \rightarrow q)$ and $p \land \sim q$
22. $\sim(p \Leftrightarrow q)$ and $(p \lor q) \land (\sim p \lor \sim q)$
23. $(p \rightarrow q) \land r$ and $(\sim p \lor q) \land r$
24. $(p \rightarrow q) \lor r$ and $\sim p \lor q \lor r$
25. $\sim[(p \rightarrow q) \land r]$ and $(p \land \sim q) \lor \sim r$

2.5 SWITCHING NETWORKS

Networks with wires and switches of the type discussed at the end of the Section 2.4 are examples of *switching networks*. As a first application of the ideas and methods introduced in the preceding sections, we shall show how to analyze and design such networks. By way of motivation you can consider the following problem:

Example 1
Three switches, P, Q, and R, are to control a single light in such a way that throwing any switch will turn the light on when it is off, and off when it is on (see Figure 2.8). How are the switches to be wired? This problem will be solved at the end of this section.

The simplest type of a switching network is given in Figure 2.9. Current flows between the terminals if and only if there is an unbroken

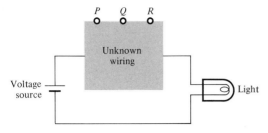

FIGURE 2.8

path between them; here, as in succeeding diagrams, we omit reference to the voltage source. Thus, there is no current flow in Figure 2.9(a) because the switch is open, but the closed switch in Figure 2.9(b) provides an unbroken path between the terminals, thereby permitting a flow of current.

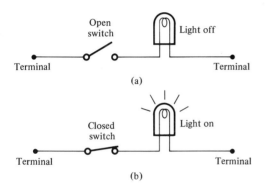

FIGURE 2.9   *Switching network*

In the discussion below we agree to the following:

(1)   Current flows between any two points in a switching network if and only if there is an unbroken path between them.
(2)   A switch is either open or closed. An open switch corresponds to a break in the network, and no current flows across it.
(3)   Switches will be designated with roman capital letters P, Q, R, and so on.
(4)   For any switch P, P' will stand for a switch which is open when P is closed, and closed when P is open.
(5)   *p* will stand for the statement "switch P is closed," *q* for "switch Q is closed," and so on.

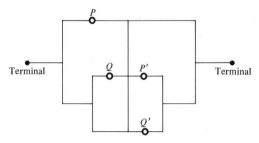

FIGURE 2.10

Example 2
Can current flow between the terminals in Figure 2.10?

SOLUTION
    The terminals are joined by an unbroken path which does not depend on the positions of P and Q (Figure 2.11). Hence, current always flows between the terminals and the switches are completely redundant. This network is equivalent to a simple wire connecting the terminals.

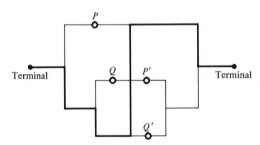

FIGURE 2.11

Example 3
    Is current flowing between the terminals in Figure 2.12 if switches P and Q are closed?

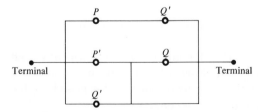

FIGURE 2.12

SOLUTION

Since P and Q are closed, the switches P' and Q' are open. Hence, no current flows across them, and as a consequence no current flows between the terminals. Observe, however, that current will flow if P or Q are open.

In the following example we show how switching networks can be related to compound statements.

Example 4

Consider the switching network in Figure 2.13. Current will flow between the terminals T_1 and T_2 when P is closed, Q is closed, or both P and Q are closed. Hence, the statement "current is flowing" is true if and only if the statement "$p \lor q$" is true. This tells us that the truth table of $p \lor q$ gives a complete analysis of the switching network in Figure 2.13 (see Table 2.20).

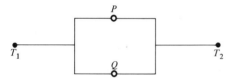

FIGURE 2.13

TABLE 2.20 DETERMINING THE CURRENT FLOW FOR THE NETWORK IN FIGURE 2.13

P	Q	Current is flowing		
Closed	Closed	T	T	T
Closed	Open	T	T	F
Open	Closed	T	F	T
Open	Open	F	F	F
		$p \lor q$	p	q

We observe that there is a dual relationship between Figure 2.13 and the statement $p \lor q$. Namely, the wiring diagram can be regarded as a schematic diagram of this statement.

To analyze switching networks effectively, we must be able to

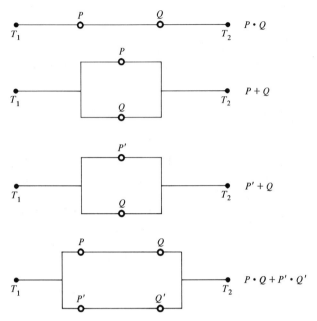

FIGURE 2.14   *Basic switching networks*

describe them symbolically. In order to accomplish this, we have to
express them in terms of simple "building blocks." All this is done in
Figure 2.14, where we introduce four basic switching networks and
their symbolic representation.

The switches in the network $P \cdot Q$ are said to be wired *in series*;
the switches in the network $P + Q$ are said to be wired *in parallel*. A
little deliberation shows that current flow in the basic switching networks is governed by Table 2.21.

TABLE 2.21   CURRENT FLOW IN THE BASIC NETWORKS
COMPARED WITH THE TRUTH TABLES OF
THE CORRESPONDING COMPOUND STATEMENTS

P	Q	$P \cdot Q$	$P + Q$	$P' + Q$	$P \cdot Q + P' \cdot Q'$		
Closed	Closed	T	T	T	T	T	T
Closed	Open	F	T	F	F	T	F
Open	Closed	F	T	T	F	F	T
Open	Open	F	F	T	T	F	F
		$p \wedge q$	$p \vee q$	$\backsim p \vee q$	$(p \wedge q) \vee (\backsim p \wedge \backsim q)$	p	q

---

By virtue of Table 2.21 we can state the following: To determine the current flow in a given switching network:

(1) Describe the network in symbolic form.
(2) Make the identification

$$\cdot \quad \text{and} \quad \wedge$$
$$+ \quad \text{and} \quad \vee$$
$$' \quad \text{and} \quad \sim$$

(3) Obtain the corresponding compound statement and find the current flow from its truth table.

We point out that

(1) $\sim p \vee q$ is equivalent to $p \rightarrow q$ (Example 3, Section 2.4).
(2) $(p \wedge q) \vee (\sim p \wedge \sim q)$ is equivalent to $p \Leftrightarrow q$ (verify with truth tables).

Hence, each of the basic connectives can be realized as a switching network.

Example 5

Determine the current flow in the switching network described in Figure 2.15.

SOLUTION

This network can be described symbolically as

$$P + [Q' + (P' \cdot Q)]$$

Hence, the current flow is read off from the truth table of

$$p \vee [\sim q \vee (\sim p \wedge q)]$$

This is done in Table 2.22. Observe that current flows between the terminals for any position of the switches.

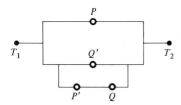

FIGURE 2.15

TABLE 2.22 CURRENT FLOW IN THE NETWORK IN FIGURE 2.15

p	q	P	+	[Q'	+	(P'	·	Q)]	P	Q
		p	∨	[~q	∨	(~p	∧	q)]		
T	T	T	T	F	F	F	F	T	Closed	Closed
T	F	T	T	T	T	F	F	F	Closed	Open
F	T	F	T	F	T	T	T	T	Open	Closed
F	F	F	T	T	T	T	F	F	Open	Open

Example 6

Determine the current flow in the switching network

$[P \cdot Q + P'] \cdot Q'.$

SOLUTION

The compound statement corresponding to this network is $[(p \wedge q) \vee \sim p] \wedge \sim q$, and the current flow is therefore given in Table 2.23.

TABLE 2.23 CURRENT FLOW IN THE NETWORK $[P \cdot Q + P'] \cdot Q'$

p	q	[P · Q + P'] · Q'					P	Q
		[(p ∧ q)	∨	~p]	∧	~q		
T	T	T	T	F	F	F	Closed	Closed
T	F	F	F	F	F	T	Closed	Open
F	T	F	T	T	F	F	Open	Closed
F	F	F	T	T	T	T	Open	Open

Remark

Observe that in constructing Table 2.23 we departed from past procedure by entering the truth values of $p \wedge q$ directly in Step 1. By now, also the reader should be able to do that.

We observe that current flows through the network only when both switches P and Q are open. The network is sketched in Figure 2.16.

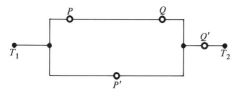

FIGURE 2.16 *The switching network* $[\mathrm{P} \cdot \mathrm{Q} + \mathrm{P}'] \cdot \mathrm{Q}'$

SOLUTION OF EXAMPLE 1

Listed in Table 2.24 are the two schemes in which throwing any switch will turn the light on when it is off, and off when it is on. We find, for example, that if we use Scheme 2, then the light will be on when the statement

$$(p \wedge q \wedge r) \vee (\sim p \wedge \sim q \wedge r) \vee (\sim p \wedge q \wedge \sim r) \vee (p \wedge \sim q \wedge \sim r)$$

is true, and off when it is false. We therefore have the switching network

$$P \cdot Q \cdot R + P \cdot Q \cdot R' + P \cdot Q' \cdot R + P' \cdot Q \cdot R$$

TABLE 2.24 THE POSSIBLE SCHEMES FOR EXAMPLE 1

Position of switches			Light	
p	q	r	Scheme 1	Scheme 2
●T	T	T	Off	On
T	T	F	On	Off
T	F	T	On	Off
F	T	T	On	Off
●F	F	T	Off	On
●F	T	F	Off	On
●T	F	F	Off	On
F	F	F	On	Off

This network consists of four parallel arrays, each with three switches (see Figure 2.17).

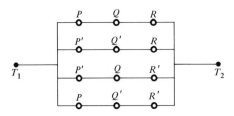

FIGURE 2.17   *Solution of Example 1*

Note:

All switches in Fig. 2.17 with the same letter or the same letter with a prime are assumed to be connected in such a way that they are on together or off together. This remark applies also to previously discussed switching networks.

EXERCISES

Find the compound statements corresponding to the switching networks in Exercises 1–8.

1. $P + Q + R$ **Answer** $p \lor q \lor r$
2. $P \cdot Q \cdot R$
3. $P \cdot (Q + R)$
4. $P + Q \cdot R$
5. $P \cdot Q' + Q' \cdot P$
6. $P \cdot [Q \cdot (R + R')]$
7. $P \cdot R + Q \cdot R$
8. $P \cdot Q + P \cdot R + Q \cdot R + P' \cdot Q' \cdot R'$

Find the switching network corresponding to the compound statements in Exercises 9–16.

9. $q \rightarrow p$ **Answer** $Q' + P$
10. $p \land q \rightarrow r$
11. $p \rightarrow (p \lor r)$
12. $p \Leftrightarrow \sim p$
13. $(p \lor \sim q) \land (\sim p \lor q)$
14. $[(p \land q) \lor (p \land r)] \rightarrow \sim r$
15. $(\sim p \land \sim q) \Leftrightarrow (p \lor q)$
16. $(p \rightarrow q) \rightarrow (p \Leftrightarrow r)$

17. Give the second switching network which solves Example 1.

In Exercises 18–25 determine if current flows in each of the switching networks if P, Q, and R are *closed*.

18. $P \cdot Q' + P \cdot R' + Q \cdot R'$
19. $(P + Q + R) \cdot (P' + Q')$
20. $(P + Q) \cdot [R + (P' \cdot Q')]$
21. $P \cdot Q' + P' \cdot Q + R'$
22. $P \cdot Q \cdot R + P' + Q' + R'$
23. $P' \cdot Q + R' \cdot Q + P \cdot Q' + P \cdot Q$
24. $(P' + Q' + R') \cdot (P + Q + R) + P$
25. $(P + Q) \cdot (P' + Q) \cdot (P + Q')$

26. Sketch diagrams for the switching networks in the given exercises:

(a)	Exercises 1, 2	(f)	Exercise 8
(b)	Exercise 3	(g)	Exercise 18
(c)	Exercise 5	(h)	Exercise 19
(d)	Exercise 6	(i)	Exercise 23
(e)	Exercise 7	(j)	Exercise 24

2.6 VALID ARGUMENTS

The second application of our newly acquired tools is aimed at the problem of drawing conclusions from given premises. In the discussion that follows we shall use the term *argument* for the claim that certain statements, called *premises*, imply a given statement, called *conclusion*.

● **Example 1**

Consider the argument

If the light is on, then the switch is on. } premise
The light is on. } premise
Therefore the switch is on. } conclusion

To decide if this conclusion is warranted, let us first reduce the argument to symbolic form. Letting p stand for "the light is on," q for "the switch is on," we have

$$\begin{matrix} p \to q \\ p \end{matrix} \Bigg\} \text{premises}$$
$$\overline{\qquad\qquad\qquad}$$
$$\therefore q \quad \} \text{conclusion} \tag{1}$$

where the symbol $\therefore$ stands for *therefore*. We now look at the truth table of the statement

$$[(p \to q) \land p] \to q$$

TABLE 2.25   THE TRUTH TABLE OF $[(p \rightarrow q) \wedge p] \rightarrow q$

p	q	$[(p \rightarrow q)$ Premise	$\wedge$	$p]$ Premise	$\rightarrow$	q Conclusion	
T	T	T	T	T	T	T	← Premises and conclusion are true
T	F	F	F	T	T	F	
F	T	T	F	F	T	T	
F	F	T	F	F	T	F	

that is, we wish to know when the conjunction of the premises $p \rightarrow q$ and p implies the conclusion q (see Table 2.25). We observe that when the premises are true (only the first row in the table), then the conclusion is also true. When this is the case we say that the argument is *valid*:

DEFINITION

An argument is *valid* if and only if the conjunction of its premises implies the conclusion.

Thus, in a valid argument, true premises always imply a true conclusion, but when one or more premises are false, then the conclusion may have any truth value.

Example 2

Consider the following argument: Investigations of the pollution of a river serving factories A, B, and C, resulted in the ensuing reports:

> A or B is responsible.
> A is responsible or C is not.
> B or C is responsible.   } premises
> If A is responsible, then B is responsible.
> B is responsible and C is not.
> Therefore B is responsible.   } conclusion

Is this argument valid?

SOLUTION

Let a stand for "A is responsible," b for "B is responsible," and C for "C is responsible." Then our argument has the symbolic representation

$$
\left.
\begin{array}{l}
a \lor b \\
a \lor \sim c \\
b \lor c \\
a \to b \\
b \land \sim c
\end{array}
\right\} \text{ premises}
$$

$$
\therefore b \qquad \left.\right\} \text{ conclusion}
$$

To solve this problem, we construct the truth table of each premise and the conclusion (Table 2.26). From this table we see that the conclusion is true whenever all the premises are true. The argument is therefore valid.

It is important to point out that we are not concerned with the truth of the premises but only with the validity (truth) of the arguments. You should therefore not be surprised to see that a true conclusion can be implied by premises which cannot possibly be true.

TABLE 2.26 DECIDING THE VALIDITY OF THE ARGUMENT IN EXAMPLE 2

a	b	c	$a \lor b$	$a \lor \sim c$	$b \lor c$	$a \to b$	$b \land \sim c$	b	
T	T	T	T	T	T	T	F	T	
T	F	T	T	T	T	F	F	F	
F	T	T	T	F	T	T	F	T	
F	F	T	F	F	T	T	F	F	
T	T	F	T	T	T	T	T	T	⌐Premises and
T	F	F	T	T	F	F	F	F	conclusion are
F	T	F	T	T	T	T	T	T	⌐true
F	F	F	F	T	F	T	F	F	

Example 3

Consider the argument

If the sun revolves around the earth, then $2 + 3 = 5$.
The sun revolves around the earth.
Therefore, $2 + 3 = 5$.

Letting p stand for "the sun revolves around the earth," q for "$2 + 3 = 5$," we can put our argument in the symbolic form

$$p \rightarrow q$$
$$\underline{p}$$
$$\therefore q$$

From Example 1 we already know that this argument is valid. As facts, the conclusion q is true, yet the premise p is false.

Example 4
Is the following argument valid?

Automobiles with high-performance engines and gasolines with a high-octane rating greatly contribute to air pollution.
Automobiles with high-performance engines greatly contribute to air pollution.
Therefore, gasolines with a high-octane rating do not greatly contribute to air pollution.

SOLUTION
This argument can be put in the symbolic form

$$p \wedge q$$
$$\underline{p}$$
$$\therefore \sim q$$

Since the statements $(p \wedge q) \wedge p$ and $p \wedge q$ are equivalent (i.e., they have the same truth table), also the statements $[(p \wedge q) \wedge p] \rightarrow \sim q$ and $(p \wedge q) \rightarrow \sim q$ are equivalent. From Table 2.27 we see, therefore, that our argument is not valid because the conjunction of the premises does not imply the conclusion (i.e., it is not a tautology). Specifically, it is the first row in the table which makes the statement invalid.

TABLE 2.27   THE TRUTH TABLE OF $(p \wedge q) \rightarrow \sim q$

p	q	$p \wedge q$	$\rightarrow$	$\sim q$
T	T	T	F	F
T	F	F	T	T
F	T	F	T	F
F	F	F	T	T

It may be worthwhile to emphasize that to decide the validity of an argument, you have to check if the conclusion is true for each case in which all the premises are true. To show that an argument is invalid, you need find only one instance in which all premises are true but the conclusion is false.

Example 5

Is the following argument valid for real numbers A and B?

If $A < \frac{1}{2}$ and $B < \frac{1}{2}$, then $A + B < 1$.
$A + B < 1$.
Therefore, $A < \frac{1}{2}$ or $B < \frac{1}{2}$.

SOLUTION

Let us write this statement in symbolic form by letting p stand for "$A < \frac{1}{2}$," q for "$B < \frac{1}{2}$," and r for "$A + B < 1$." This gives

$$(p \land q) \rightarrow r$$
$$r$$
$$\overline{}$$
$$\therefore p \lor q$$

From Table 2.28 we deduce that when p and q are false and r is true, then the conjunction of the premises $[(p \land q) \rightarrow r] \land r$ is true, whereas the conclusion $p \lor q$ is false. Hence, $\{[(p \land q) \rightarrow r] \land r\} \rightarrow (p \lor q)$ is not a tautology, and by definition the argument is not valid.

TABLE 2.28 DECIDING THE VALIDITY OF THE ARGUMENT IN EXAMPLE 3

p	q	r	$(p \land q) \rightarrow r$	r	$p \lor q$
T	T	T	T	T	T
T	F	T	T	T	T
F	T	T	T	T	T
F	F	T	T	T	F←
T	T	F	F	F	T
T	F	F	T	F	T
F	T	F	T	F	T
F	F	F	T	F	F

Valid arguments and switching networks

Switching networks can be used to determine the validity of arguments as follows: Consider, for example, the argument

$$p \to q$$
$$\underline{p }$$
$$\therefore q$$

(See Example 1.) This argument is true because the statement

$$[(p \to q) \wedge p] \to q \tag{2}$$

is a tautology. Now this statement is equivalent to

$$\sim[(p \to q) \wedge p] \vee q$$

by Exercise 19 of Section 2.4. This statement is simplified as follows:

> $(p \to q) \wedge p$ is equivalent to $(\sim p \vee q) \wedge p$ (again Example 5, Section 2.4)
> $\sim[(\sim p \vee q) \wedge p]$ is equivalent to $\sim(\sim p \vee q) \vee \sim p$ (Exercise 19, Section 2.3)
> $\sim(\sim p \vee q) \vee \sim p$ is equivalent to $(p \wedge \sim q) \vee \sim p$ (Exercise 22, Section 2.3)

Hence, statement (2) is equivalent to

$$(p \wedge \sim q) \vee \sim p \vee q \tag{3}$$

Associate p with the statement "switch P is closed," q with "switch Q is closed." Then (2) is realized by means of the switching network in Figure 2.18. Current will flow through this network regardless of the position of the switches P and Q, and hence the light will always be on. In a switching network associated with an inavlid argument, the light will be off for certain positions of the switches. Hence, the

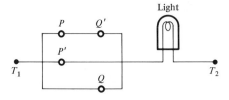

FIGURE 2.18

validity of an argument can be decided by checking out the appropriate switching network. Such an operation can be carried out automatically with computers.

EXERCISES

Put the arguments in Exercises 1–8 into symbolic form and decide their validity.

1. If A and B are real numbers, and $A \cdot B = 0$, then $A = 0$ or $B = 0$.
   $A = 0$.
   Therefore, $B \neq 0$.

2. If the box is not empty, it contains ten balls.
   The box is empty.
   Therefore, the box does not contain ten balls.

3. Set A is empty or set B is empty.
   If set A is empty, then set C is empty.
   If set C is empty, then set B is empty.
   Therefore, sets A and B are empty.

4. The computer is protected if fuses A and B are installed.
   Fuse A is installed.
   Therefore, the computer is not unprotected.

5. The light is on if and only if the switch is on.
   The switch is off.
   Therefore, the light is off.

6. Consider real numbers A and B:

   If A is irrational, then $1 + A$ is irrational.
   If B is irrational, then $1 + B$ is irrational.
   Therefore, $1 + A$ and $1 + B$ are irrational.

7. Sea water can be used for drinking if and only if it is desalted.
   If appropriate research is not carried out, then sea water cannot be desalted.
   Appropriate research is carried out.
   Therefore, sea water can be used for drinking.

8. If it does not rain, then the game will take place this Saturday.
    If the game takes place this Saturday, then the picnic takes
      place on Sunday.
    It does not rain.
    Therefore, the picnic takes place on Sunday.

Decide the validity of the arguments in Exercises 9–12.

9. $p \Leftrightarrow q$
    $q \Leftrightarrow r$
    $\therefore p \Leftrightarrow r$

10. $p \lor q$
    $p \land r$
    $\therefore q \lor r$

11. $p \to \sim q$
    $q \to p$
    $\therefore p \land q$

12. $p \lor q$
    $p \to q$
    $\sim q$
    $\therefore \sim p$

In Exercises 13–16 draw diagrams for the switching networks which
correspond to the given arguments.

13. $p \lor q$
    $\sim p$
    $\therefore \quad q$

14. p
    q
    $\therefore p \lor q$

15. $p \to q$
    $q \to r$
    $\therefore p \to r$

16. p
    $\sim q$
    $\therefore \sim(p \land q)$

3
COUNTING

3

3.1 THE NUMBER OF ELEMENTS IN A SET

In this and the next chapter we shall be concerned with a variety of counting problems. The counting process assigns a unique positive integer $n(A)$ to each finite set A.

DEFINITION 1

A set A is *finite* if

(1) A is the empty set, and then $n(A) = 0$, or
(2) the members of A can be matched in a one-to-one manner with a set of positive integers $\{1, 2, 3, \ldots, m\}$ for some m, and then $n(A) = m$.

All sets considered will be assumed to be finite.

Example 1

If $A = \{2, 4, 6, 8, 10\}$, then $n(A) = 5$.
If $A = \{a, b, c, d, e, f, g\}$, then $n(A) = 7$.

We begin with the following principle:

If A and B are finite sets such that $A \cap B = \emptyset$, then

$$n(A \cup B) = n(A) + n(B) \qquad (1)$$

The truth of this theorem can be seen intuitively (see Figure 3.1). If $n(A) = m$ and $n(B) = k$, then the set A can be matched with

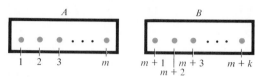

FIGURE 3.1

the set $\{1, 2, 3, \ldots, m\}$, and the set B can be matched with the set $\{m + 1, m + 2, \ldots, m + k\}$. This matches the set $A \cup B$ with the set $\{1, 2, \ldots, m, m + 1, \ldots, m + k\}$, and hence $n(A \cup B) = m + k$.

Formula (1) is no longer true when the sets A and B are not disjoint, as shown in the following example.

Example 2
If $A = \{a, b, c, d, e, f\}$ and $B = \{d, e, f, g, h\}$, then

$$A \cup B = \{a, b, c, d, e, f, g, h\}$$

Hence, $n(A) = 6$, $n(B) = 5$, and $n(A \cup B) = 8$, so that

$$n(A \cup B) \neq n(A) + n(B)$$

We observe, however, that A and B have three elements in common and that $8 = 6 + 5 - 3$. This suggests the formula

$$n(A \cup B) = n(A) + n(B) - n(A \cap B)$$

This formula generalizes Formula (1), and it is true in general that

---

If A and B are any finite sets, then

$$n(A \cup B) = n(A) + n(B) - n(A \cap B) \tag{2}$$

There are formulas also for the number of elements in a union of more than two sets, but it is easier and more instructive to work out such problems directly.

Example 3
In a laboratory experiment, a number of rats had to pass through mazes A and B. One observer counted 28 rats passing through A, a second observer counted 12 rats passing through B, and a third observer counted 7 rats passing through both A and B. How many rats participated in the experiment?

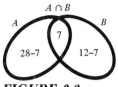

FIGURE 3.2

SOLUTION

Consult Figure 3.2. We have $n(A) = 28$, $n(B) = 12$, and $n(A \cap B) = 7$. By Formula (2), therefore,

$$n(A \cup B) = 28 + 12 - 7 = 33$$

The total number of participating rats was 33.

Example 4

At the end of a busy day a salesman found entered in his note-book the following orders for writing pads:

Stores a and b:	70 dozen
Stores a and c:	40 dozen
Store a:	120 dozen
Store b:	100 dozen
Store c:	75 dozen

What is the actual total of the orders?

SOLUTION

This type of problem is best solved using a Venn diagram (see Figure 3.3). The entries are entered from the center out: The circled numbers indicate the step at which this part of the diagram was filled

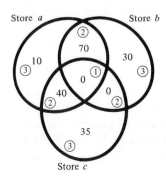

FIGURE 3.3

in. Thus, in step 3 we take into account that of the 120 dozen attributed to a, 110 dozen were already accounted for in step 2, and so on. The total order is seen to be

$$10 + 30 + 35 + 70 + 40 = 185 \text{ dozen}$$

We conclude this section with a counting principle which will be of central importance throughout this text. We motivate it with an example.

Example 5

Suppose you have the following three tests for an initial classification of rocks:

Color test   (white-gray, yellow, brown-red, black, other)
Luster test   (dull, shiny)
Hardness test   (hard, soft)

How many different types of rock can be classified with these tests?

SOLUTION

We begin with the observation that the result of a given test does not depend on the other two tests and neither does it influence their outcome. That is, the tests are independent of one another. Hence, for each of the five color types we can perform two luster tests, and this gives a total of $5 \times 2 = 10$ possible types of rock. Each of these possible types can be further distinguished by using the hardness test. Hence, the total number of types of rock that can be classified with the three tests is $5 \times 2 \times 2 = 20$.

The following general principle is valid for any finite sequence of events with a finite number of possible outcomes. An *event* as used here may mean an experiment, answering a question on an examination, a game, and so on.

Fundamental counting principle

Consider a sequence of n events $P, Q, R \cdots$. Suppose that event P has p possible outcomes; suppose that for each of these, event Q has q possible outcomes; suppose that for each pair of possible outcomes of P and Q, the event R has r possible outcomes, and so on. Then the number of possible outcomes for the sequence $P, Q, R, \cdots$ is the product of the outcomes $p \times q \times r \times \cdots$.

In Example 5 the events are color test (P), luster test (Q), and hardness test (R), and the respective number of possible outcomes is $p = 5$, $q = 2$, and $r = 2$.

EXERCISES

1. Of n students in a mathematics class, 45 are economics majors, 72 are mathematics majors, and 28 are double majors. Find n.

2. Grounds for divorce in some states are the following:

   Pregnancy at marriage (P)
   Nonsupport (N)
   Mental cruelty (M)

   A survey of k states revealed the following information on grounds for divorce:

   3 states, P
   6 states, N
   5 states, M
   2 states, P and N
   2 states, P and M
   4 states, N and M
   2 states, P and N and M
   1 state, none of these

   Find k.

3. A survey of three groups of adults revealed the following information:

   Group A: of 19 members, 4 are members of B, 7 are members of C
   Group B has 11 members
   Group C has 15 members

   What is the total membership of the three groups?

4. Among the set of positive integers not exceeding 100, let A be the set of integers divisible by 5, B the set of integers divisible by 10. Find $n(A \cup B)$.

5. To provide for adequate customer service and personnel to handle inventory, a department store has among its employees: 120

salesmen, 42 cashiers, 68 stockroom workers. Among these, however,

36 are salesmen and stockroom workers
30 are cashiers and stockroom workers
5 are salesmen, cashiers, and stockroom workers

How many employees does this make up?

6. To settle claims, an insurance company sent out the following checks:

171 checks in January
102 checks in February
 88 checks in March

An audit showed, however, that the following errors occurred: 59 checks mailed in February and 47 checks mailed in March duplicated checks mailed in January; 12 of the checks mailed each month duplicated 12 checks in each of the two other mailings. What was the total number of claims actually settled?

7. Consider the set $S = \{1, 2, 3, \ldots, 30\}$ and the following subsets.

$A = \{x | x \text{ is even}\}$
$B = \{x | x \text{ is divisible by } 3\}$
$C = \{x | x \text{ is divisible by } 5\}$

Find

(a) $n(A \cup B)$
(b) $n(B \cup C)$
(c) $n(A \cup B \cup C)$

8. What is $n(A \cup B \cup C)$ if

$n(A) = 17 \quad n(A \cap B) = 5 \quad n(A \cap B \cap C) = 2$
$n(B) = 19 \quad n(A \cap C) = 2$
$n(C) = 21 \quad n(B \cap C) = 5$

9. What relations must exist between the sets A, B, and C if

(a) $n(A \cup B \cup C) = n(A) + n(B) + n(C)$
(b) $n(A \cup B \cup C) = n(A \cap B \cap C)$
(c) $n(A \cup B \cup C) = n(A) + n(B) - n(A \cap B)$

10.   Find a formula similar to Formula (2) for $n(A \cup B \cup C)$.
11.   Consider a ten-question multiple-choice test with three possible answers for each question. How many different answers are there for the test?
12.   A three-place combination lock has ten digits for each place. How many different combinations are possible?

3.2   PERMUTATIONS—THE FACTORIAL NOTATION

Among the most fascinating and elusive problems in mathematics and its applications are so-called counting problems. The following are examples of such problems.

Example 1
   One key in a set of ten will open a lock. In how many ways can the keys be inserted in the lock in sequence? In how many cases will the key which opens the lock be the last to be tried?

Example 2
   In how many ways can a committee of five members be selected from a list of ten candidates?

   In principle, this type of problem can be solved by the method of exhaustion, that is, by listing all possibilities and counting them. Practically, this is impossible because the number of possibilities is usually much too big. The problems in Examples 1 and 2 will be solved later in this section. To solve them, we have to know something about arrangements (permutations), and to introduce this concept we ask the following question:

Example 3
   In how many ways can the letters a, b, c be placed in three boxes?

SOLUTION
   The number of possibilities is small enough to be exhausted by trial. In Figure 3.4 we note that there are six arrangements, called from now on *permutations*.

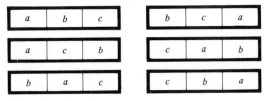

FIGURE 3.4   *The permutations on three letters*

DEFINITION 1

A *permutation* of *k* different objects is an arrangement of these objects in a definite order.

Example 4

How many permutations are there of the letters *a*, *b*, *c*, *d?*

SOLUTION

We shall also solve Example 4 by listing all possible permutations, but we shall use a scheme which will give us some insight into this type of problem. The diagram in Figure 3.5 is an example of what we shall call a *tree*. A path through the tree from 0 to finish will be called a *branch*. We see that each branch determines a permutation, and conversely, each permutation determines a branch.

An inspection of Figure 3.5 will show that there is a total of 24 permutations on 4 distinct objects. We also notice that there are 4

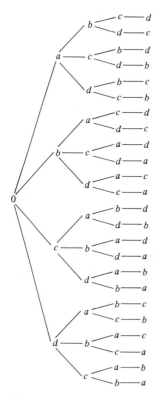

FIGURE 3.5

first choices, 3 second choices for each first choice, 2 third choices for each second choice, and 1 last choice for each third choice; that is,

$$24 = 4 \times 3 \times 2 \times 1$$

We see that for the permutations on three distinct objects, we have

$$6 = 3 \times 2 \times 1$$

Reflecting on these two examples, we suspect that the following important principle is true:

THEOREM 1

There is a total of $n \times (n - 1) \times (n - 2) \times \cdots \times 3 \times 2 \times 1$ permutations on a set of n distinct objects.

The truth of this theorem is made intuitively clear when we imagine that a tree like that in Figure 3.5 is drawn for n objects instead of four objects. We now introduce new notation for designating the number of permutations on distinct objects and work several more examples.

DEFINITION 2

The product of all integers from 1 to n, inclusive, will be designated with the symbol $n!$, called "*n-factorial*":

$$n! = n \times (n - 1) \times \cdots \times 3 \times 2 \times 1$$

Example 5

$$
\begin{aligned}
1! &= 1 \\
2! &= 2 \times 1 = 2 \\
3! &= 3 \times 2 \times 1 = 6 \\
4! &= 4 \times 3 \times 2 \times 1 = 24 \\
5! &= 5 \times 4 \times 3 \times 2 \times 1 = 120 \\
6! &= 6 \times 5 \times 4 \times 3 \times 2 \times 1 = 720 \\
7! &= 7 \times 6 \times 5 \times 4 \times 3 \times 2 \times 1 = 5,040 \\
8! &= 8 \times 7 \times 6 \times 5 \times 4 \times 3 \times 2 \times 1 = 40,320 \\
9! &= 9 \times 8 \times 7 \times 6 \times 5 \times 4 \times 3 \times 2 \times 1 = 362,880 \\
10! &= 10 \times 9 \times 8 \times 7 \times 6 \times 5 \times 4 \times 3 \times 2 \times 1 = 3,628,800
\end{aligned}
$$

Observe that the numbers $n!$ increase very rapidly with increasing n. The number 5! has 3 digits, 10! has 7 digits, and 100! has 158 digits.

We are now ready to solve Example 1, restated here for your convenience.

Example 6
One key in a set of ten will open a lock. In how many ways can the keys be inserted in the lock in sequence? In how many cases will the key which opens the lock be the last to be tried?

SOLUTION

The first part of the theorem asks for the count of the number of sequences of ten trials in which each key is used once. Each sequence of trials is seen to be simply a permutation on the set of keys. Hence, the number of ways in which the keys can be inserted is

$$10! = 3,628,800$$

If we consider only those sequences of trials in which one particular key is always last, then the count will be the number of permutations on the remaining nine keys (see Figure 3.6). Hence, there are $9! = 362,880$ cases in which the correct key is tried last.

Nine wrong keys   Correct key

FIGURE 3.6

Next, we solve Example 2, restated here for convenience.

Example 7
In how many different ways can five committee assignments be made from a list of ten candidates?

SOLUTION

Suppose one of the ten candidates has been given an assignment. A second assignment can then be made from the remaining nine candidates, given $10 \times 9 = 90$ possibilities for the two assignments. A third assignment can now be made from the remaining eight candidates, and so on. The number of ways in which five committee assignments can be made from the ten candidates is seen to be

$$10 \times 9 \times 8 \times 7 \times 6 = 30,240$$

We note in passing that this number can be written as

$$10 \times 9 \times 8 \times 7 \times 6$$
$$= \frac{10 \times 9 \times 8 \times 7 \times 6 \times 5 \times 4 \times 3 \times 2 \times 1}{5 \times 4 \times 3 \times 2 \times 1} = \frac{10!}{5!}$$

From Example 7 we see that the following general principle is true:

THEOREM 2

If a selection can be made in k ways and, after any one of these selections is made, another can be made in m ways, and then a third selection can be made in n ways, and so on, then the total number of successive selections is $k \times m \times n \times \cdots$.

The following is a more sophisticated version of Example 7.

Example 8

Five different committee assignments can be made from a list of ten candidates. Among the candidates is a married couple of which only one partner can be selected. In how many ways can the committee assignments be made?

SOLUTION

Suppose no partner of the married couple has been selected. Then there are only eight candidates for the five assignments, giving $8 \times 7 \times 6 \times 5 \times 4 = 6{,}720$ possible selections. If one partner of the married couple has been selected, then there are eight eligible candidates left for the remaining four assignments, giving $8 \times 7 \times 6 \times 5 = 1{,}680$ possible selections. Since there are that many for each of the two partners, the total number of ways in which the five committee assignments can be made is

$$2 \times 1{,}680 + 6{,}720 = 10{,}080$$

EXERCISES

1. Check your understanding of the factorial notation with the following True-False questions. Circle your answer.

(a) $20! = 4! \times 5!$ T   F
(b) $5! > 2^5$ T   F
(c) $n! > 2^n$ for $n > 3$ T   F
(d) $(m + n)! = m! + n!$ T   F

(e) $n(n-1) \cdots (n-m+1) = \dfrac{n!}{m!}$ T   F

(f) $k! = k \times (k-1)!$ T   F
(g) $(m \times n)! \leq m! \times n!$ T   F
(h) $(2n)! = 2 \times n!$ T   F

2.   Referring to Example 6, in how many cases will the first six keys *not* open the lock?
3.   Referring to Example 6, in how many cases will the first or second key open the lock?
4.   In how many ways can 12 girls be dated by 12 boys? (Each girl and each boy are to have a date.)
5.   After 250 personal income tax notices were sealed in addressed envelopes, it was discovered that, due to an error, some notices were sealed in the wrong envelope.

  (a)   In how many ways could each of the notices be put into the correct envelope?
  (b)   In how many ways could it happen that exactly one notice got into the wrong envelope?
  (c)   In how many ways could it happen that all but two notices got into the right envelope?

6.   How many car license plates can be formed with three distinct digits followed by three distinct letters? (The digits are 0, 1, . . ., 9, and the letters are taken from the 26 letters of the alphabet.)
7.   From his home in Hollywood, a student has a choice of seven routes leading to the bus stop, and then three different bus routes to UCLA. How many choices does he have for a round trip from his home to UCLA and back?
8.   How many six-digit numbers can be made with the digits 0, 1, 2, . . ., 9 if no digit is repeated? (No number can begin with 0.)
9.   How many six-digit numbers can be made up from the digits 1, 2, . . ., 9 if the digit 1 is to appear twice but no other digit is repeated?
10.   How many nine-digit numbers can be formed from the digits 1, 2, . . ., 9 if even and odd digits must alternate and no digit is repeated?
11.   How many nine-digit numbers can be formed from the digits 1, 2, . . ., 9 if the first and last digits must have the same parity (both are odd or both are even) and no digit is repeated?
12.   What is the smallest number of nonzero digits needed to form 7! seven-digit numbers if no digit is used twice?

13. In how many ways can 28 students be lined up for a group picture?

14. In how many ways can 28 students be lined up for a group picture if 2 particular students are to be in front, and 26 in back?

15. In how many ways can 28 students be lined up for a picture if a particular student must stand at the extreme left?

16. There are 14 female and 14 male students. In how many ways can they be lined up for a picture if

   (a) Female students occupy the front row and male students occupy the back row?

   (b) The students are lined up in such a way that males and females alternate?

17. A future astronaut wishes to chart a course that will pass once each of the other eight planets of our solar system. How many routes does he have to choose from?

18. Referring to Exercise 17, how many routes are there if Mars must be the fifth planet visited?

19. Draw a tree diagram for the permutations on the three letters *a*, *b*, *c*:

   (a) For all permutations.

   (b) When the first letter cannot be *a*.

   (c) When the last letter must be *a*.

3.3 CYCLIC PERMUTATIONS AND PERMUTATIONS WITH REPETITION

To explain what is meant by a cyclic permutation, consider the following problem.

Example 1

In how many ways can three keys be put on a key ring?

SOLUTION

   In a problem such as this it is always assumed that two permutations are different only if they cannot be made to coincide with a rotation. In Figure 3.7 you will observe that the three situations in (a) can be made to coincide with a rotation, as can those in (b). This is to say that the relative position of the keys is the same in each of the cases (a) and (b). Thus, to find the number of permutations in which the relative order of the keys is different, we can pick any key as an "observer" and count the permutations of the remaining keys

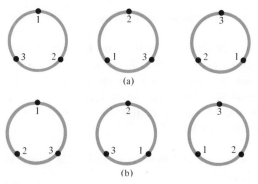

(a)

(b)

FIGURE 3.7

relative to it. Hence, there are $2 \times 1 = 2$ distinct ways in which the keys can be put on the key ring.

Reflecting on Example 1, the following general principle is seen to be true.

THEOREM 1

There are $(n - 1)!$ cyclic permutations on n distinct objects.

Example 2

A drug manufacturer uses the digits 1, 2, 3, 4, 5, 6, 7, printed around the bottom on the label of a small round container to identify the supplier, the date of manufacture, and the plant. Will this be sufficient to cover a three-year period if the manufacturer has four suppliers and seven plants, each producing one batch per month?

SOLUTION

Over the three-year period, each plant will produce 36 batches which have to be identified. The number of different identifications the manufacturer needs is therefore $4 \times 7 \times 36 = 1,008$. On the other hand, the number of cyclic permutations on seven different objects is $6! = 720$. Hence, the seven digits will not do.

Example 3

At a board meeting, ten persons, including the chairman and his secretary, are seated around a circular table. In how many ways can this be done if the chairman and his secretary must sit next to each other?

SOLUTION

There are only two ways in which the secretary can be seated relative to the chairman, but for each of these there are 8! permutations on the remaining eight persons. Hence, there are $2 \times 8! = 80{,}640$ different seating arrangements. Note that without the restriction on the seating of the chairman and his secretary, there would have been $9! = 362{,}880$ different seating arrangements.

In the preceding examples we dealt with permutations on distinct (distinguishable) objects. How is the answer affected when some of the objects in a permutation are alike? This will be illustrated in the following examples.

Example 4
How many distinct permutations are there on the four letters a, a, a, b?

SOLUTION

A moment's reflection will convince you that a complete listing of all permutations is

$$
\begin{array}{cccc}
a & a & a & b \\
a & a & b & a \\
a & b & a & a \\
b & a & a & a \\
\end{array}
$$

How can we arrive at the answer 4 without listing all possibilities? We observe that a given permutation is not changed by permuting the indistinguishable letters a among themselves. There are $3!$ such "indistinguishable permutations" on the three letters a, a, a, but a total of $4!$ permutations on four letters. Thus, if P is the number of different permutations, then $P \times 3! = 4!$, so that

$$P = \frac{4!}{3!} = 4$$

The tree of all distinct permutations is given in Figure 3.8.

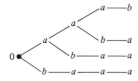

FIGURE 3.8

Example 5
How many distinct permutations are there on the letters a, a, a, b, b?

SOLUTION

Listing all possible distinct permutations is not an easy matter here, so we argue as follows: A given permutation remains unchanged when the three letters *a* are permuted among themselves (3! ways) and the two letters *b* are permuted among themselves (2! ways). On five letters there is a total of 5! permutations, so if the number of distinct permutations is *P*, then $P \times 3! \times 2! = 5!$, giving for the answer

$$P = \frac{5!}{3! \times 2!} = 10$$

The tree of all distinct permutations is given in Figure 3.9.

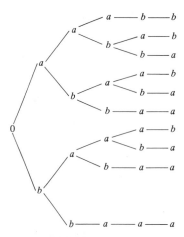

FIGURE 3.9

Example 6

How many distinct permutations are there on a deck of 52 cards if we distinguish between cards only according to color?

SOLUTION

For any of the 52! permutations on the 52 cards, there are 26! indistinguishable permutations on the red cards and 26! indistinguishable permutations on the black cards. Thus, if *P* is the number of distinct permutations, then $P \times 26! \times 26! = 52!$. Hence, the number of distinct permutations is

$$P = \frac{52!}{26!\,26!}$$

The following general principle will be discussed in Section 3.4.

THEOREM 2

Suppose there are $n = n_1 + n_2 + \cdots + n_k$ objects of which exactly n_1 are alike, n_2 are alike, and so on. Then the number P of distinct permutations is

$$P = \frac{n!}{n_1! \, n_2! \, n_3! \cdots n_k!}$$

Example 7

How many distinct permutations are there on a deck of 52 cards if we do not distinguish between cards of the same suit?

SOLUTION

A deck of playing cards has 4 suits of 13 cards each. Hence,

$$P = \frac{52!}{13! \, 13! \, 13! \, 13!} = \frac{52!}{(13!)^4}$$

EXERCISES

1. In how many ways can ten (different) keys be put on a key ring?
2. In how many ways can ten keys be put on a key ring if three particular keys are to be kept together in the same order?
3. Four black beads and one white bead are used for a necklace. How many different necklaces can be made if

   (a) The black beads are alike?
   (b) The black beads are of different sizes?

4. How many different bracelets can be made with 12 beads if

   (a) Eleven beads are alike and one is different?
   (b) All 12 beads are different?
   (c) Four beads each are yellow, orange, and red, but except for color they are alike, and beads of the same color are to be kept together?

5. How many different permutations are there on the letters a, a, b, b, c, c, d?
6. How many different permutations are there on the digits of the number 121121112?
7. How many distinct permutations on the letters in STATISTICS are there?

8. In how many ways can a 20-question True-False test be answered if it is known that exactly half the questions are true?
9. How many two-digit numbers are there?
10. How many four-digit numbers are there?
11. A class of 98 students is to be divided into 7 sections of 14 students each. In how many ways can this be done?
12. A department-store manager has ten employees to staff three departments with three persons each and to select a floor manager. In how many ways can this be done?

3.4 SUBSETS OF GIVEN SET—COMBINATIONS

We now wish to consider the type of problem related to subsets of a given set.

Example 1

For a study, a subset of 10 persons is to be selected from a population X of 30 persons. How many different subsets of 10 members can be selected?

SOLUTION

We begin with the observation that the order of members *within* a selected set is irrelevant. To find the number of subsets of 10 members that can be selected from a set of 30 members, we argue as follows: Imagine the members of X to be lined up in some order, and select the first 10 for the study. Call this set S (see Figure 3.10). This set S will not change if we perform on X those 10! permutations involving only members of S. Likewise, S will not change if we perform on X those 20! permutations involving no member of S. Every permutation on X which interchanges members of S with members not in S gives a new subset of 10 members. There is a total of 30! permuta-

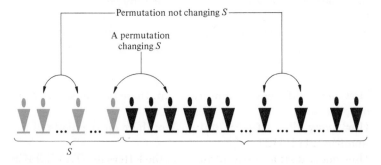

FIGURE 3.10

tions on X, and if P is the number of permutations giving different sets S, then

$$P \times 10! \times 20! = 30!$$

and hence

$$P = \frac{30!}{10! \times 20!} \tag{3}$$

It is needless to say that the number in (3) is too large to work out without a calculator ($P = 30{,}045{,}015$), and therefore the answer should be left in the given form. A comparison with Section 3.3 shows that (3) also answers the following problem.

Example 2

Let a set X have 30 objects of two kinds: 10 identical objects of one kind and 20 identical objects of another kind. What is the number of distinct permutations on X?

By way of generalization, if we set

$$30 = n \quad \text{and} \quad 10 = m$$

then $20 = n - m$, and Formula (3) becomes

$$P = \frac{n!}{m!(n-m)!}$$

Upon reflection, the following result will be seen to be true:

A set of n elements has $\dfrac{n!}{m!(n-m)!}$ different subsets of m elements.

We shall state this fact in a better form after we introduce some notation. The numbers $\dfrac{n!}{m!(n-m)!}$ appear so often that they are designated with special symbol $\dbinom{n}{m}$:

$$\binom{n}{m} = \frac{n!}{m!(n-m)!} \tag{4}$$

The numbers $\binom{n}{m}$ are called *binomial coefficients*; the reason for this will become clear in Section 3.5.

Example 3

$$\binom{7}{5} = \frac{7!}{5! \times 2!} = \frac{7 \times 6 \times \cancel{5!}}{\cancel{5!} \times 2} = \frac{7 \times 6}{2} = 21$$

$$\binom{7}{2} = \frac{7!}{2! \times 5!} = 21$$

$$\binom{8}{1} = \frac{8!}{1! \times 7!} = \frac{8 \times \cancel{7!}}{\cancel{7!}} = 8$$

$$\binom{22}{20} = \frac{22!}{20! \times 2!} = \frac{22 \times 21 \times \cancel{20!}}{\cancel{20!} \times 2} = \frac{22 \times 21}{2} = 231$$

In case $n = m$, Formula (4) becomes

$$\binom{n}{n} = \frac{n!}{n! \times (n - n)!} = \frac{n!}{n! \times 0!}$$

This expression is undefined because the sumbol $0!$ is undefined. It is convenient to have a meaning for this symbol, and we define

$$0! = 1$$

With this, Formula (4) is defined for all integers $m = 0, 1, 2, \ldots, n$. In particular, we note that

$$\binom{n}{0} = \frac{n!}{0! \times (n - 0)!} = \frac{n!}{n!} = 1$$

$$\binom{n}{1} = \frac{n!}{1! \times (n - 1)!} = \frac{n \times \cancel{(n-1)!}}{\cancel{(n-1)!}} = n$$

$$\binom{n}{n - 1} = \frac{n!}{(n - 1)! \times 1!} = n$$

$$\binom{n}{n} = \frac{n!}{n! \times (n - n)!} = \frac{n!}{n!} = 1$$

We can now state the following fact.

THEOREM 1

A set of n elements has $\binom{n}{m}$ different subsets of m elements.

In particular, $\binom{n}{0} = 1$ gives the number of subsets having 0 elements (the empty set), $\binom{n}{1} = n$ gives the number of singletons, and $\binom{n}{n} = 1$ gives the number of subsets having n elements. *Observe that the empty set and the set itself are counted as subsets.* Theorem 1 deals with a set of n elements and its subsets of m elements. Order does not count here, and in this context a subset of m elements is often called an "m-combination" of n elements taken m at a time. Problems such as those discussed in this section are called *combination problems*.

Example 5

Evaluate $\binom{4}{m}$ for $m = 0, 1, 2, 3, 4$.

SOLUTION

$$\binom{4}{0} = 1$$

$$\binom{4}{1} = 4$$

$$\binom{4}{2} = \frac{4!}{2!\,2!} = \frac{4 \times 3 \times \cancel{2!}}{2 \times \cancel{2!}} = \frac{4 \times 3}{2} = 6$$

$$\binom{4}{3} = \frac{4!}{3!\,1!} = \frac{4 \times 3!}{3!} = 4$$

$$\binom{4}{4} = 1$$

Example 5
How many samples of 4 items are there in a lot of 50 items?

SOLUTION

This problem asks for the number of combinations of 50 elements taken 4 at a time. The answer is

$$\binom{50}{4} = \frac{50!}{4! \times 46!} = \frac{50 \times 49 \times 48 \times 47 \times \cancel{46!}}{4 \times 3 \times 2 \times 1 \times \cancel{46!}}$$

$$= \frac{50 \times 49 \times 48 \times 47}{4 \times 3 \times 2} = 230{,}300$$

Example 6

A shipment of 30 cameras contains one defective sample. How many batches of 4 cameras could contain the defective sample?

SOLUTION

The defective camera, together with 3 other cameras taken from the remaining 29, will make up a batch. Hence, the answer is

$$\binom{29}{3} = \frac{29!}{3! \times 26!} = \frac{29 \times 28 \times 27 \times \cancel{26!}}{3 \times 2 \times \cancel{26!}} = \frac{29 \times 28 \times 27}{6}$$

$$= 3{,}654$$

Example 7

How many hands of 5 cards from a deck of 52 playing cards contain 3 hearts?

SOLUTION

Consult Figure 3.11. Of the 52 cards of a deck, 13 are hearts. Our hand consists of two components which are independent of each other: the 3 hearts and the 2 non-hearts. The number of sets of 3 hearts is

$$\binom{13}{3} = 286$$

FIGURE 3.11

and the number of sets of 2 other cards is

$$\binom{39}{2} = 741$$

According to the fundamental counting principle stated in Section 3.1, the answer is the product

$$\binom{13}{3} \times \binom{39}{2} = 211,926$$

Example 8

How many hands of 13 cards contain 5 hearts, 4 clubs, and 3 spades?

SOLUTION

Arguing as in Example 7, the hearts can be selected in $\binom{13}{5}$ ways, the clubs in $\binom{13}{4}$ ways, and the spades in $\binom{13}{3}$ ways. The remaining card, which must be diamond, can be selected in $\binom{13}{1}$ ways. Again we obtain the answer by using the fundamental counting principle of Section 3.1. The answer is

$$\binom{13}{5} \times \binom{13}{4} \times \binom{13}{3} \times \binom{13}{1}$$

Example 9

How many diagonals does a hexagon have?

SOLUTION

A hexagon has six vertices. Any two vertices determine a side or a diagonal. The number of combination of vertices taken two at a time gives the number of diagonals plus sides; this number is

$$\binom{6}{2} = \frac{6!}{2! \times 4!} = 15$$

Since there are six sides, the number of diagonals is $15 - 6 = 9$.

EXERCISES

1.   Evaluate the following expressions (binomial coefficients).

(a) $\binom{5}{m}$ for $m = 0, 1, 2, 3, 4, 5$

(b) $\binom{6}{m}$ for $m = 0, 1, 2, \ldots, 6$

(c) $\binom{101}{100}$

(d) $\binom{100}{3}$

(e) $\binom{85}{2} - \binom{85}{83}$

(f) $\binom{90}{0}$

(g) $\binom{54}{54}$

2.   The following is a list of True-False questions. Circle your answer.

(a) $\binom{4}{1} + \binom{5}{2} = \binom{6}{2}$ T   F

(b) $\binom{12}{0} + \binom{12}{1} = \binom{13}{1}$ T   F

(c) $\binom{24}{12} = \binom{2}{1}$ T   F

(d) $\binom{n}{m} = \binom{n+1}{m+1}$ T   F

(e) $\binom{n}{1} = \binom{n}{n-1}$ T   F

(f) $\binom{4+5}{3} = \binom{4}{3} + \binom{5}{3}$ T   F

3.   For quality control, a transistor manufacturer tests 5 transistors in every batch of 10,000. In how many ways can the 5 transistors be selected from a batch of 10,000?

4.   A set of 10 women is to be selected from a population of 70 women and 65 men. In how many ways can this be done?

5. A set of 10 men and 10 women is to be selected from a population of 70 women and 60 men. In how many ways can this be done?

6. A shipment of 1,000 television sets contains 17 defective sets.

    (a) In how many ways can a sample of 20 sets be selected from the shipment?
    (b) How many of the sample sets contain no defective television set?
    (c) How many of the sample sets contain 15 defective television sets?

7. A 20-question True-False test has eight True answers. In how many ways can the eight True answers be assigned?

8. From a semester's assignment of 100 problems, the instructor selects 20 for the final examination. In how many ways can this be done?

9. How many hands of 5 cards are there of the same suit?

10. How many hands of 13 cards contain 3 aces and 3 kings?

11. How many hands of 13 cards contain exactly 1 heart, 1 spade, and 1 club?

12. How many hands of 5 cards contain 4 aces?

13. How many hands of 13 cards contain no club?

14. How many diagonals does an octagon have?

15. A regular n-sided polygon has n vertices. How many diagonals does it have?

16. How many combinations of six letters of the English alphabet contain no vowel?

17. How many combinations of five letters of the English alphabet contain two vowels?

18. An automobile assembly plant receives ball bearings in lots of 10,000. To check for quality control, the plant checks a random sample of 100 ball bearings from each lot. The entire lot is rejected when a sample of 100 contains 10 or more defective ball bearings.

    (a) How many samples of 100 ball bearings are there per lot?
    (b) If the lot contains 80 defective ball bearings, how many samples of 100 may contain 10 defective ball bearings?
    (c) How many samples of 100 can be free of defective ball bearings if the lot contains 80 defective ball bearings?

19. Referring to Exercise 18, how many possible samples will cause the lot to be rejected?

3.5 THE BINOMIAL THEOREM

In Section 3.4 we introduced the numbers

$$\binom{n}{m} = \frac{n!}{m!(n-m)!}$$

which are defined for each positive integer n and $m = 0, 1, 2, \ldots, n$. We have learned that, for each value of m, $\binom{n}{m}$ gives the number of subsets of m elements that can be selected from a set of n elements. In particular, $\binom{n}{0} = 1$ corresponds to the set having 0 elements (the empty set), $\binom{n}{1} = n$ gives the number of singletons, and $\binom{n}{n} = 1$ corresponds to the set having all n elements. These numbers will now be connected with a very important theorem.

Many problems in mathematics involve raising a "binomial" $(a + b)$ to a positive integral power, that is, expanding (multiplying out) $(a + b)^n$ where n is some positive integer. Our goal is to find a formula that works for all n, and this will be done through the expansions of $(x + 1)^n$.

Example 1
The following expansions are obtained by direct multiplication.

$$\begin{aligned}
x + 1 \ &= x + 1 \\
(x + 1)^2 &= x^2 + 2x + 1 \\
(x + 1)^3 &= x^3 + 3x^2 + 3x + 1 \\
(x + 1)^4 &= x^4 + 4x^3 + 6x^2 + 4x + 1 \\
(x + 1)^5 &= x^5 + 5x^4 + 10x^3 + 10x^2 + 5x + 1 \\
(x + 1)^6 &= x^6 + 6x^5 + 15x^4 + 20x^3 + 15x^2 + 6x + 1 \\
(x + 1)^7 &= x^7 + 7x^6 + 21x^5 + 35x^4 + 35x^3 + 21x^2 + 7x + 1
\end{aligned}$$

Examining these expansions, we observe that for any n, the expansion of $(x + 1)^n$ will have the form

$$(x + 1)^n = \boxed{1}x^n + \boxed{n}x^{n-1} + \boxed{}x^{n-2} + \boxed{}x^{n-3} + \cdots$$
$$+ \boxed{n}x + \boxed{1}$$

To determine this expansion, we have to put the correct coefficients into the boxes. In Figure 3.12, called Pascal's triangle, we listed the coefficients of the expansions in Example 1. The reader may be surprised to observe the symmetry of the coefficients in each row, and may be more surprised to discover that each row after the first is ob-

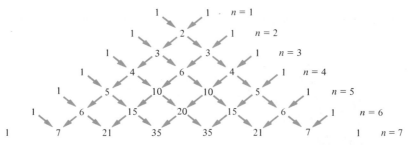

FIGURE 3.12 *Pascal's triangle*

tained from the preceding one by adding successive pairs of coefficients as indicated.

Pascal's triangle is not practical for large values of n because the nth row cannot be computed without first computing *all* preceding rows. The alert student will notice from Section 3.4, Example 4, that the fourth row in Pascal's triangle is obtained by evaluating $\binom{4}{m}$ for $m = 0, 1, 2, 3, 4$. From Exercise 1 (a) and (b) of Section 3.4, one sees that the fifth and sixth rows are computed by evaluating the expressions $\binom{5}{m}$ and $\binom{6}{m}$, respectively. At this point we should be willing to make the following assertion:

For each n, the nth row in Pascal's triangle is obtained by computing the numbers $\binom{n}{m}$ for $m = 0, 1, 2, \ldots, n$.

Example 2
Express the binomial expansion of $(x + 1)^4$ in terms of the numbers $\binom{4}{m}$.

SOLUTION

$$(x + 1)^4 = \binom{4}{0}x^4 + \binom{4}{1}x^3 + \binom{4}{2}x^2 + \binom{4}{3}x + \binom{4}{4}$$

Now, if the assertion we just made is true, then the nth row in Pascal's triangle is

$$\binom{n}{0} \quad \binom{n}{1} \quad \binom{n}{2} \quad \binom{n}{3} \cdots \binom{n}{n-3} \quad \binom{n}{n-2} \quad \binom{n}{n-1} \quad \binom{n}{n}$$

Because of the symmetry of each row in Pascal's triangle, we must have

$$\binom{n}{0} = \binom{n}{n}$$

$$\binom{n}{1} = \binom{n}{n-1}$$

$$\binom{n}{2} = \binom{n}{n-2}$$

and, in general,

$$\binom{n}{m} = \binom{n}{n-m} \quad \text{for} \quad m = 0, 1, 2, \ldots, n$$

This is, indeed true:

THEOREM 1

$$\binom{n}{n-m} = \binom{n}{m} \tag{5}$$

PROOF

By definition,

$$\binom{n}{k} = \frac{n!}{k!(n-k)!}$$

Setting $k = n - m$ gives $n - k = m$, and hence

$$\binom{n}{n-m} = \frac{n!}{(n-m)!\,m!} = \frac{n!}{m!(n-m)!} = \binom{n}{m}$$

Putting together what we have just discovered leads us to the following important result:

The binomial theorem

If n is a positive integer, then

$$(x + 1)^n = \binom{n}{0}x^n + \binom{n}{1}x^{n-1} + \binom{n}{2}x^{n-2}$$

$$+ \cdots + \binom{n}{n-1}x + \binom{n}{n} \tag{6}$$

Example 3
Find the expansion of $(x + 1)^8$.

SOLUTION

$$\binom{8}{0} = \binom{8}{8} = 1$$

$$\binom{8}{1} = \binom{8}{7} = 8$$

$$\binom{8}{2} = \binom{8}{6} = \frac{8!}{2!\,6!} = 28$$

$$\binom{8}{3} = \binom{8}{5} = \frac{8!}{3!\,5!} = 56$$

$$\binom{8}{4} = \frac{8!}{4!\,4!} = 70$$

Using Formula (6) with $n = 8$ gives

$$(x + 1)^8 = x^8 + 8x^7 + 28x^6 + 56x^5 + 70x^4$$
$$+ 56x^3 + 28x^2 + 8x + 1$$

The following fact is worth noting: In the expansion of $(x + 1)^n$, the powers x^m and x^{n-m} have the same coefficient $\binom{n}{m}$.

As an application of the binomial theorem we prove the following fact.

THEOREM 2
The number of all possible subsets of a set of n elements is 2^n.

PROOF

The number of all subsets having m elements is $\binom{n}{m}$. Hence, the number of *all* possible subsets is

$$\binom{n}{0} + \binom{n}{1} + \binom{n}{2} + \binom{n}{3} + \cdots + \binom{n}{n}$$

Using the binomial theorem with $x = 1$ gives

$$2^n = (1 + 1)^n = \binom{n}{0} + \binom{n}{1} + \binom{n}{2} + \cdots + \binom{n}{n}$$

Hence, the theorem is true.

The expansion of $(a + b)^n$

From Formula (6) we can obtain a formula for the expansion of $(a + b)^n$. The case $b = 0$ is not interesting, since $(a + 0)^n = a^n$. If $b \neq 0$, then we can substitute $x = a/b$ in Formula (6). Because

$$\left(\frac{a}{b} + 1\right)^n = \left(\frac{a + b}{b}\right)^n = \frac{(a + b)^n}{b^n}$$

we have

$$\frac{(a + b)^n}{b^n} = \binom{n}{0}\left(\frac{a}{b}\right)^n + \binom{n}{1}\left(\frac{a}{b}\right)^{n-1} + \binom{n}{2}\left(\frac{a}{b}\right)^{n-2}$$

$$+ \cdots + \binom{n}{n-1}\left(\frac{a}{b}\right) + \binom{n}{n}$$

Multiplying both sides by b^n and noting that $b^n(a/b)^{n-m} = a^{n-m}b^m$ gives

---

$$(a + b)^n = \binom{n}{0}a^n + \binom{n}{1}a^{n-1}b + \binom{n}{2}a^{n-2}b^2$$

$$+ \cdots + \binom{n}{n-1}ab^{n-1} + \binom{n}{n}b^n \quad (7)$$

Example 4
Find the expansion of $(x - y)^3$.

SOLUTION

Write $(x - y) = x + (-y)$ and use Formula (7) with $a = x$, $b = -y$, and $n = 3$:

$$(x - y)^3 = \binom{3}{0}x^3 + \binom{3}{1}x^2(-y) + \binom{3}{2}x(-y)^2 + \binom{3}{3}(-y)^3$$

$$= x^3 - 3x^2y + 3xy^2 - y^3$$

We now consider an application of Formula (7) to a type of problem in which coin tossing is one of its variations. If we write out the expantion of $(T + H)^3$ without simplifying, we get

$$(T + H)^3 = TTT + (TTH + THT + HTT)$$

$$+ (THH + HTH + HHT) + HHH \quad (8)$$

Thus, the 2^3 terms are as follows:

> 1 term with 3 T's and 0 H's
> 3 terms with 2 T's and 1 H
> 3 terms with 1 T and 2 H's
> 1 term with 0 T's and 3 H's

This fact has the following interpretation.

Example 5
If a coin is tossed three times there are the following 2^3 possible outcomes (H = Head, T = Tail):

$$\text{TTT,   (TTH, THT, HTT),   (THH, HTH, HTT),   HHH}$$

When we compare this with Formula (8), we see that the binomial coefficients, which are 1, 3, 3, and 1 in this case, give the number of all possible outcomes for each given number of Heads and Tails, as follows:

> 1 possible outcome with 3 Tails and 0 Heads
> 3 possible outcomes with 2 Tails and 1 Head
> 3 possible outcomes with 1 Tail and 2 Heads
> 1 possible outcome with 0 Tails and 3 Heads

Example 6
A coin is tossed six times. How many possible outcomes have two Heads and four Tails?

SOLUTION
The number of outcomes is the binomial coefficient of T^4H^2 in the expansion of $(T + H)^6$. This coefficient is

$$\binom{6}{4} = \frac{6!}{4!\,2!} = 15$$

Example 7
A coin is tossed eight times. In how many possible outcomes does the number of Heads exceed the number of Tails?

SOLUTION
A term H^mT^{8-m} in the expansion of $(H + T)^8$ corresponds to a possible outcome with m Heads and $8 - m$ Tails. Hence, we are inter-

ested here in the binomial coefficients of the terms H^8, H^7T, H^6T^2, and H^5T^3. The number of possible outcomes with Heads exceeding Tails is therefore

$$\binom{8}{8} + \binom{8}{7} + \binom{8}{6} + \binom{8}{5} = \binom{8}{0} + \binom{8}{1} + \binom{8}{2} + \binom{8}{3}$$

$$= 1 + 8 + 28 + 56 = 93$$

EXERCISES

Expand the expressions in Exercises 1–10 and simplify.

1. $(x + y)^4$ 2. $(x + y)^5$
3. $(x + y)^6$ 4. $(x - y)^4$
5. $(x - y)^5$ 6. $(x - y)^6$
7. $(2x + y)^6$ 8. $(x + x^2)^4$

9. $\left(x + \dfrac{1}{x}\right)^5$ 10. $(1 + 3x)^3$

Evaluate the numbers in Exercises 11–14.

11. $(1 + \frac{1}{4})^4$ 12. $(1.1)^4$
13. $(1.01)^4$ 14. $(2.01)^4$

Hint:
$(1.1)^4 = (1 + 0.1)^4$

15. In the expansion of $(x + y)^{40}$, find the following coefficients:

   (a) $\square\, x^3 y^{37}$
   (b) $\square\, x^{20} y^{20}$
   (c) $\square\, x^{10} y^{30}$

16. In the expansion of $(x + 2y)^{30}$, find the following coefficients:

   (a) $\square\, x y^{29}$
   (b) $\square\, x^{29} y$
   (c) $\square\, x^{15} y^{15}$

17. How many subsets of three or fewer elements does a set of ten elements have?
18. How many subsets of eight or more elements does a set of nine elements have?

19. Show that

$$\binom{n}{0} - \binom{n}{1} + \binom{n}{2} - \binom{n}{3} + \cdots + (-1)^n\binom{n}{n} = 0$$

20. Show that $\binom{n+1}{m} = \binom{n}{m-1} + \binom{n}{m}$

Hint:

Express each symbol in terms of factorials, find the common denominator for the right side, and add.

21. In how many ways can a 20-question True-False test be answered?

22. A coin is tossed 12 times.

(a) In how many possible outcomes are there as many Heads as Tails?

(b) In how many possible outcomes are there exactly 8 Tails?

(c) In how many possible outcomes are there exactly 4 Heads or exactly 4 Tails?

3.6 PARTITIONS

The problem of classification plays an important part in any area which involves a systematic study of objects or events. Examples of such areas are medical diagnosis, disease control, insecticides, government studies, grammar, life sciences, and others. The concept of partition is an important application of set theory to this problem.

A partition of a set is a division of the set into disjoint subsets.

DEFINITION 1

A collection of sets $[A_1, A_2, \ldots, A_r]$ is a *partition* of a set X if two conditions hold:

(1) $X = A_1 \cup A_2 \cup \cdots \cup A_r$

(2) If $i \neq j$, then $A_i \cap A_j = \emptyset$

The subsets A_i are called the *cells* of the partition.

Example 1

Let $X = \{a, b, c, d, e, f\}$. The following are partitions of X:

$[\{a, b\}, \{c, d, e, f\}]$

$[\{a, b, c\}, \{d, e, f\}]$

$[\{a, b\}, \{c, d\}, \{e, f\}]$

The collection of subsets

$$[\{a, b\}, \{c, d, e\}, \{e, f\}]$$

is *not* a partition because $\{c, d, e\} \cap \{e, f\} \neq \emptyset$

The refinement which makes the concept of partition useful is the *cross-partition*. We introduce it in the following example:

Example 2

A botanist prepares a rudimentary classification of California wild flowers into two categories:

Region: [desert (R_1), mountains (R_2), shore (R_3)]
Color: [white to pink (C_1), rose to purplish red (C_2), blue to violet (C_3), yellow to orange (C_4), other (C_5)]

Each of the two categories is a partition of the set of wild flowers. A classification superior to each separate category is obtained when we consider all subsets $R_i \cap C_j$, $i = 1, 2, 3$, $j = 1, 2, 3, 4, 5$. The collection of these subsets gives a partition called the *cross-partition* of $[R_1, R_2, R_3]$ and $[C_1, C_2, C_3, C_4, C_5]$. It is schematically presented in Figure 3.13. Observe that the cross-partition consists of $3 \times 5 = 15$ cells.

	C_1	C_2	C_3	C_4	C_5
R_1	$C_1 \cap R_1$	$C_2 \cap R_1$	$C_3 \cap R_1$	$C_4 \cap R_1$	$C_5 \cap R_1$
R_2	$C_1 \cap R_2$	$C_2 \cap R_2$	$C_3 \cap R_2$	$C_4 \cap R_2$	$C_5 \cap R_2$
R_3	$C_1 \cap R_3$	$C_2 \cap R_3$	$C_3 \cap R_3$	$C_4 \cap R_3$	$C_5 \cap R_3$

FIGURE 3.13

DEFINITION 2

If $[A_1, A_2, \ldots, A_r]$ and $[B_1, B_2, \ldots, B_s]$ are two partitions of a set X, then the collection of cells

$$C_{ij} = A_i \cap B_j \qquad i = 1, 2, \ldots, r \qquad j = 1, 2, \ldots, s$$

is called the *cross-partition* of $[A_1, A_2, \ldots, A_r]$ and $[B_1, B_2, \ldots, B_s]$.

Example 3

Consider the set $X = \{1, 2, 3, 4, 5, 6, 7, 8\}$ and subsets

$A_1 = \{1, 3, 5, 7\}$ $B_1 = \{1, 3, 5, 6, 7\}$
$A_2 = \{2, 4, 6, 8\}$ $B_2 = \{2, 4, 8\}$

Then $[A_1, A_2]$ and $[B_1, B_2]$ are different partitions of X. If we put

$$C_{11} = A_1 \cap B_1 = \{1, 3, 5, 7\}$$
$$C_{12} = A_1 \cap B_2 = \emptyset$$
$$C_{21} = A_2 \cap B_1 = \{6\}$$
$$C_{22} = A_2 \cap B_2 = \{2, 4, 8\}$$

then

$$[C_{11}, \emptyset, C_{21}, C_{22}] = [\{1, 3, 5, 7\}, \emptyset, \{6\}, \{2, 4, 8\}]$$

is the cross-partition of $[A_1, A_2]$ and $[B_1, B_2]$. Note that C_{12} has been listed, even though $C_{12} = \emptyset$.

Intuitively, a cross-partition of two partitions of a set is itself a partition of the set. A proof of this fact is given at the end of the section; it may be omitted without losing the continuity of the material.

Counting partitions

We are often interested in partitions in which the cells are given in a specific order, and these are called *ordered partitions*.

Example 4

Let $X = \{a, b, c, d\}$. Then

$$[\{a, b\}, \{c, d\}] \quad \text{and} \quad [\{c, d\}, \{a, b\}]$$

are two, distinct, ordered partitions of the set X.

Let us derive a formula for the number of ordered partitions of a set (a formula for counting unordered partitions will also be derived). Suppose a set X of $n = n_1 + n_2 + \cdots + n_r$ elements is partitioned into r cells, a cell with n_1 elements, a cell with n_2 elements, and so on. In how many ways can this be done if the cells are given in a specific order?

Consider a given ordered partition of X. This partition does not change when we permute the elements in each cell among themselves, but the partition changes when we permute elements between cells. Hence, the number of ordered partitions of X is the number of permutations on X if the elements of each cell are regarded to be alike. By Theorem 2 of Section 3.3, this number is

$$\frac{n!}{n_1! \, n_2! \cdots n_r!}$$

The notation

$$\binom{n}{n_1,\ n_2,\ \cdots,\ n_r} = \frac{n!}{n_1!\ n_2!\ \cdots\ n_r!} \tag{9}$$

is often used to designate the number of ordered partitions. Since there are $r!$ permutations on the r cells of the ordered partition, the number of unordered partitions is

$$\frac{1}{r!}\binom{n}{n_1,\ n_2,\ \cdots,\ n_r}$$

Example 4

Evaluate $\binom{10}{1,\ 2,\ 7}$.

SOLUTION

$$\binom{10}{1,\ 2,\ 7} = \frac{10!}{1!\ 2!\ 7!} = \frac{10 \times 9 \times 8 \times \cancel{7!}}{2 \times \cancel{7!}} = \frac{10 \times 9 \times 8}{2} = 360$$

Observe that for the case $r = 2$,

$$\binom{n}{n_1,\ n_2} = \binom{n}{n_1} = \binom{n}{n_2}$$

Example 6

A class of 27 students is to be divided into three discussion sections of equal size. In how many ways can this be done?

SOLUTION

This problem asks for the number of (unordered) partitions of a set of 27 elements into 3 cells of 9 elements each. The answer, therefore, is

$$\frac{1}{3!}\binom{27}{9,\ 9,\ 9} = \frac{1}{6} \times \frac{27!}{9!\ 9!\ 9!} = \frac{27!}{6 \times (9!)^3}$$

The answer should be left in the given form, since the numbers in the numerator and the denominator are much too large for a simplification.

Dichotomies

The partition of a set into two cells is given the special name *dichotomy*. As a special case of Formula (9) we have the fact that

---

the number of dichotomies of a set of n elements into a cell of m elements and a cell of $n - m$ elements is $\binom{n}{m}$.

For the following example and Exercises 10 and 11, the reader may wish to review Sections 2.2–2.4.

Example 7

In Section 2.3 we discussed the assignment of truth values to statements p and q (see Table 2.5). The set of logical possibilities for p and q can be written as

$$L = \{TT, TF, FT, FF\}$$

The logical possibilities for which the statement

$$P: (p \vee q) \wedge (\sim p \vee q)$$

is true determine a dichotomy of L. Find it.

SOLUTION

According to Table 2.10, the statement P is true for the logical possibilities $\{TT, FT\}$ and false for $\{TF, FF\}$. Hence, the dichotomy is

$$[\{TT, FT\}, \{TF, FF\}].$$

We end this section with a proof of the following fact.

THEOREM 1

A cross-partition of two partitions of X is itself a partition of X.

PROOF

Let $[A_1, A_2, \ldots, A_r]$ and $[B_1, B_2, \ldots, B_s]$ be two partitions of X, and consider the cross-partition with cells $C_{ij} = A_i \cap B_j$. We must show that

(1) X is the union of cells C_{ij}
(2) The cells C_{ij} are disjoint.

(1)   Since $A_i \subset X$ and $B_j \subset X$, we also have $A_i \cap B_j \subset X$ for $i = 1, 2, \ldots, r$ and $j = 1, 2, \ldots, s$. Hence, the union of all cells C_{ij} is *contained* in X. Next, if $x \in X$, then $x \in A_i$ for some value of i and $x \in B_j$ for some value of j, since $[A_1, A_2, \ldots, A_r]$ and $[B_1, B_2, \ldots B_s]$ are partitions of X. Hence, $x \in A_i \cap B_j$, telling us that the union of cells C_{ij} *contains* X. We conclude that X actually *equals* the union of cells C_{ij}.

(2)   Consider two cells $C_{ij} = A_i \cap B_j$ and $C_{km} = A_k \cap B_m$ (see Figure 3.14). Then, by Section 1.1,

$$C_{ij} \cap C_{km} = (A_i \cap B_j) \cap (A_k \cap B_m) = (A_i \cap A_k) \cap (B_j \cap B_m)$$

If $C_{ij} \neq C_{km}$, then $i \neq k$ or $j \neq m$. In the first case, $A_i \cap A_k = \emptyset$; and in the second case, $B_j \cap B_m = \emptyset$. In either case, however, $C_{ij} \cap C_{km} = \emptyset$. Hence, the cells of the cross-partition are disjoint.

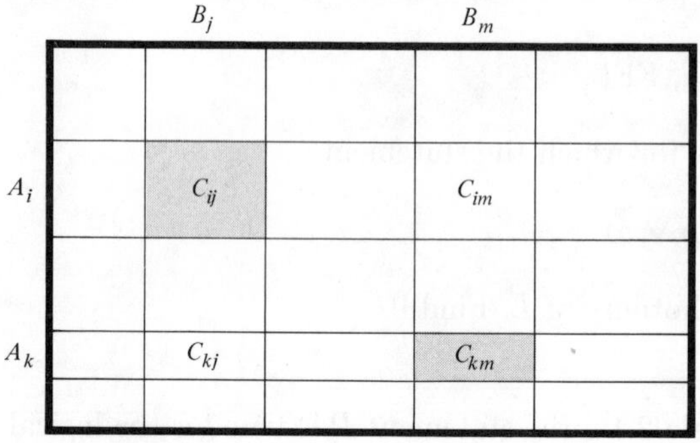

FIGURE 3.14

EXERCISES

1.   Which of the following are partitions of $X = \{a, b, c, d, e\}$?

(a) $[\{a\}, \{b\}, \{c, d, e\}]$
(b) $[\{a, b, c\}, \{c, d, e\}]$
(c) $[\{a\}, \{b\}, \{c\}, \{d\}]$
(d) $[\{a, e\}, \{b, c\} \{d\}]$

2.   Find the cross-partition of each pair of the partitions of

$X = \{a, b, c, d, e\}.$

(a) $[\{a\}, \{b, c, d, e\}]$ $[\{a, b, c, d\}, \{e\}]$
(b) $[\{a, b\}, \{c, d, e\}]$ $[\{a, b, c\}, \{d, e\}]$
(c) $[\{a, c, e\}, \{b, d\}]$ $[\{a, b\}, \{c\}, \{d, e\}]$

3.   Compute the following numbers:

(a) $\begin{pmatrix} 20 \\ 1,\,1,\,18 \end{pmatrix}$

(b) $\begin{pmatrix} 20 \\ 1,\,0,\,19 \end{pmatrix}$

(c) $\begin{pmatrix} 20 \\ 17,\,3 \end{pmatrix}$

(d) $\begin{pmatrix} 7 \\ 4,\,2,\,1 \end{pmatrix}$

(e) $\begin{pmatrix} 40 \\ 0,\,0,\,40 \end{pmatrix}$

(f) $\begin{pmatrix} 40 \\ 2,\,36,\,2 \end{pmatrix}$

Exercises 4–7 pertain to a deck of 52 playing cards.

4.   How many partitions into four cells of 13 cards are there? How many ordered partitions?

5.   How many partitions are there into four cells of 13 cards each if one cell is to contain the four aces? How many ordered partitions?

6.   In a partition into one 12-cell and two 20-cells, one 20-cell is to contain the four aces, and the other the four kings. How many such partitions are there? How many ordered partitions?

7.   How many ordered partitions of a set of 52 elements are there into five singletons and one 47-cell?

8.   A ten-question test has six true and four false answers. In how many ways can the test be answered?

9.   Referring to Table 2.1: Replace "no" by 0 and "yes" by 1. Let $[A, B]$ be the partition of the listed states such that A contains all states for which the sum of the digits is 2 or 3, and B the other states. Find the partition.

10. Referring to Example 7: Find the dichotomy of L determined by the truth values of each of the following statements.

(a) $p \lor q$
(b) $p \land q$
(c) $p \lor \sim q$
(d) $\sim(\sim p \land \sim q)$
(e) $p \to q$ (see Table 2.15)
(f) $p \Leftrightarrow (q \to \sim p)$ (see Table 2.16)

11. Let L be the set of logical possibilities of truth values of statements pqr:

$$L = \{\text{TTT, TTF, TFT, FTT, FFT, FTF, TFF, FFF}\}$$

(see Table 2.6). Find the dichotomy of L determined by the truth values of each of the following statements:

(a) $(p \wedge q) \vee r$
(b) $(p \wedge q) \vee (p \wedge r)$
(c) $(p \vee q) \wedge \sim(p \wedge r)$
(d) $[(p \wedge q) \vee (p \wedge r)] \vee \sim(p \vee r)$ (see Table 2.11)
(e) $p \Leftrightarrow (q \wedge \sim r)$ (see Table 2.19)
(f) $(p \rightarrow q) \rightarrow (q \rightarrow r)$
(g) $(p \vee q) \Leftrightarrow (p \vee r)$

3.7   APPLICATIONS

In this section we describe various applications of successive partitions and cross-partitions to identification problems. The material of this section is not crucial to the understanding of the rest of the text.

Example 1   (Successive Partitions)

Of seven optically identical steel balls, six have the same weight and one is slightly heavier. Find the odd steel ball by using a pan-balance twice.

SOLUTION

Partition the steel balls as shown in Figure 3.15(a), putting three balls into each pan. If the pans are balanced, the remaining ball is the heavier one. Otherwise, take the heavier set of balls, and partition it as shown in Figure 3.15(b). If the pans balance, the remaining ball is the heavier one. Otherwise, the scale will tell which the heavier one is.

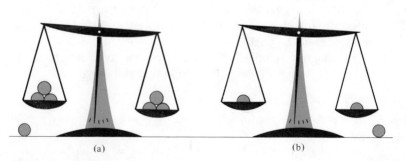

(a)   (b)

FIGURE 3.15

Example 2   (Successive Dichotomies)

In the typical "20-questions game" you must make an identification through a sequence of questions which your opponent answers yes or no. In the present game the cards in each suit of a regular deck are numbered 1 through 13. Your opponent picked a card. Identify it.

SOLUTION

The successive dichotomies are conveniently represented by means of a *tree* (Figure 3.16). At each junction you pick one of the alternatives as your question and the yes or no answer will tell you which branch to follow. For example, the following sequence of questions and answers leads to 2-heart:

Black? → no → Heart? → yes → 1–6? → yes → 1, 2, 3?

→ yes → 1? → no → 2? → yes

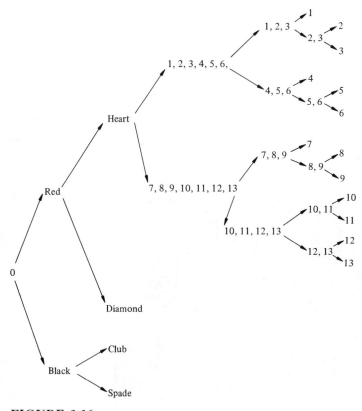

FIGURE 3.16

Example 3 (Independent Dichotomies)

Many decisions we have to make, rather than being clear-cut, call for a compromise. To narrow his search for living accommodations, a college student decided to go about it in a systematic way. He prepared two lists of desirable features. Since he wanted at least three features from each list, he entered a digit 1 in the table when a place had that feature, 0 otherwise, and added the digits in each row of each list (see Table 3.1).

Designate the set of places under consideration X:

$$X = \{a, b, c, d, e, f, g, h\}$$

The first list gives a dichotomy $[A_1, A_2]$ with cells

$$A_1 = \{a, b, c, d, e, g\} \qquad \text{(sum of digits} \geq 3)$$
$$A_2 = \{f, h\} \qquad \text{(sum of digits} < 3)$$

The second list gives a dichotomy $[B_1, B_2]$ with cells

$$B_1 = \{a, d, g, h\} \qquad \text{(sum of digits} \geq 3)$$
$$B_2 = \{b, c, e, f\} \qquad \text{(sum of digits} < 3)$$

The student's choice will be made from places in

$$A_1 \cap B_1 = \{a, d, g\}$$

TABLE 3.1

	LIST I						LIST II				
Place considered	Rent less than 120 dollars	Walking distance from campus	Shopping within three blocks	Cafeteria nearby	Bus service	SUM OF DIGITS	Laundry room	Kitchen facilities	Television	Back yard	SUM OF DIGITS
a	1	1	0	0	1	3	1	1	1	0	3
b	1	0	1	1	1	4	0	0	1	1	2
c	0	1	0	1	1	3	0	1	0	1	2
d	1	1	0	1	1	4	1	1	1	0	3
e	0	1	1	1	1	4	1	1	0	0	2
f	0	1	0	1	0	2	0	1	1	0	2
g	1	1	0	0	1	3	0	1	1	1	3
h	1	0	1	0	0	2	1	1	0	1	3

	B_1	B_2
A_1	$A_1 \cap B_1$	$A_1 \cap B_2$
A_2	$A_2 \cap B_1$	$A_2 \cap B_2$

FIGURE 3.17

The cross-partition of $[A_1, A_2]$ and $[B_1, B_2]$ is given schematically in Figure 3.17.

Remark

It may occur to the reader to ask, "Why bother with two dichotomies in the first place? Why not start with a single list of *all* features, and then consider only places for which the sum of the digits is 6 or higher?" If we did this, we would arrive at a dichotomy with cells

$$C_1 = \{a, b, d, e, g\} \qquad \text{(sum of digits} \geq 6)$$

$$C_2 = \{c, f, h\} \qquad \text{(sum of digits} < 6)$$

In this dichotomy, b and e have a high rating, but these places are not as desirable by the criteria of our student.

EXERCISES

1. Of 12 optically identical steel balls, 11 have the same weight and one is slightly lighter. What is the least number of times you have to use a pan-balance to find the lighter steel ball?

2. Of five optically identical steel balls, two weigh the same, and three weigh the same, but they are slightly heavier. What is the least number of times you have to use a pan-balance to identify the two lighter balls?

3. Consult Table 2.1. You are to identify with a sequence of questions with yes-no answers the state whose grounds for divorce are none of those listed. What is the least number of questions? List your questions.

4. What is the smallest number of yes–no questions necessary to identify any one of the 50 states?

5. How many yes-no questions are needed to identify any two digits picked from the set $\{1, 2, 3, 4, 5, 6\}$? List your questions.

6. A 1,000-page dictionary has an average of 20 words per page, listed in two columns.

(a) How many yes-no questions are necessary to get you to any page?

(b) What is the largest number of yes-no questions necessary to locate any word if it is assumed that each page has at most 15 words per column?

7. Locate a telephone directory. Estimate the number of yes-no questions will identify a given name in the directory.

8. A set of eight lights is wired in such a way that if one bulb is defective, none will light. You have four good spare bulbs. Describe a sequence of three tests that will identify the defective bulb.

9. Referring to Exercise 8: Describe the tests which will identify two defective light bulbs.

4
PROBABILITY

4

4.1 A DEFINITION OF PROBABILITY

The term *probability* is used in daily language to describe the relative possibility that an event will occur. The events are wide-ranging, and they include cancer research, gambling, election returns, weather forecast, and so on. We say "relative possibility" because when discussing the likelihood of any event, we must take into account also those possible events which can interfere with it and even prevent it. The preceding description of probability is, of course, very imprecise and incomplete, but it contains the germ of the idea, which is that probability is a measure of possibility of the occurrence of an event. This is made clearer with the following two examples.

Example 1

Consider two intersecting sets A and B. With what certainty can we assert that if $x \in A$, then $x \in B$?

For the situation depicted in Figure 4.1(a) we can assert with absolute certainty that $x \in B$, since $A \subset B$. We cannot be as certain when the situation is that in Figure 4.1(b), but we can assert with "about 70 percent" certainty that if $x \in A$, then $x \in B$. This is based on the observation that about 70 percent of the area of A lies in B. What can we assert when the situation is like that in Figure 4.1(c)?

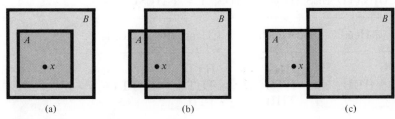

(a)   (b)   (c)

FIGURE 4.1

For x to be in B when x is in A is one of many possible *events*, and a statement like "there is a 70 percent certainty that if $x \in A$, then $x \in B$" expresses the *probability* of the event.

Example 2

The possible outcomes of three tosses of a coin are listed in Table 4.1. What are the chances that the outcome will be exactly two Heads?

SOLUTION

The number of all possible outcomes (logical possibilities) of this experiment is $2^3 = 8$. The set of possible outcomes is called a *sample space*. By a direct count we see that three possible outcomes have exactly 2 Heads. We assume that Head is as likely to come up as Tail, and a reasonable answer to the question is therefore this:

"The chances of exactly 2 Heads are 3 out of 8"

This chance is expressed as the ratio 3/8 and called the *probability* of the event "2 Heads." We express this with the abbreviated notation

$$P(2 \text{ Heads}) = \frac{3}{8}$$

Let us phrase the question using set notation. If we let S designate the sample space of possible outcomes in Table 4.1, and let

$$A = \{\text{HHT, HTH, THH}\}$$

(see Figure 4.2), then we can ask

"What are the chances that an element of S is in A?"

TABLE 4.1 THE POSSIBLE OUTCOMES OF THREE COIN TOSSES (H = HEADS, T = TAILS)

3 Heads	2 Heads	1 Head	0 Heads
HHH	HHT HTH THH	HTT THT TTH	TTT
0 Tails	1 Tail	2 Tails	3 Tails

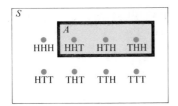

FIGURE 4.2

Using the notation of Section 3.1, we have $n(A) = 3$ and $n(S) = 8$, and hence,

$$P(2 \text{ Heads}) = \frac{n(A)}{n(S)}$$

The probability that if $x \in S$, then $x \in A$ can also be expressed as $P(x \in A)$. This is abbreviated $P(A)$, and we shall express a formula like this also in the equivalent form

$$P(A) = \frac{n(A)}{n(S)}$$

Example 3

In three tosses of a coin, what is the probability that the outcome is 3 Tails?

SOLUTION

Table 4.1 shows that there is only one outcome with 3 Tails. We therefore say that the probability is $1/8$, written

$$P(3 \text{ Tails}) = \frac{1}{8}$$

If we put $A = \{TTT\}$, then we have

$$P(A) = \frac{n(A)}{n(S)} = \frac{1}{8}$$

In the above discussion it was assumed that, in tossing a coin, Head is as likely to come up as Tail. We say that the *events* Head and Tail are *equiprobable*. We also observe that, since $A \subset S$, we have

$$0 \leq n(A) \leq n(S)$$

and hence,

$$0 \leq \frac{n(A)}{n(S)} \leq 1$$

The probabilities we considered are therefore numbers between 0 and 1. This will be a general feature of probabilities. Observing in Examples 1–3 that events can be expressed in terms of subsets of a set of possible outcomes (logical possibilities), we state:

DEFINITION 1

A *sample space* S is a set of logical possibilities of a given experiment.

The elements of S are called *sample points*.

An *event* is a subset of a sample space.

Example 4

What is the sample space when a single die is rolled?

SOLUTION

This question is ill-posed because the answer depends on the events being considered. Two common cases are these:

(1)   The dots on each face of a die represent the numbers 1 through 6. Hence a sample space can be chosen to be $S_1 = \{1, 2, 3, 4, 5, 6\}$.

(2)   Suppose that we distinguish only between odd and even numbers. The sample space can thus be $S_2 = \{\text{odd, even}\}$.

However, if $A = \{1, 3, 5\}$, then to say that $x \in S$ is odd is equivalent to saying that $x \in A$. Likewise, if $B = \{2, 4, 6\}$, then $x \in S$ is even when and only when $x \in B$. Hence, even in this case we can retain the sample space S_1 if we take A and B as the events "odd" and "even," respectively.

In general, the sample space has to be figured out from the problem at hand. We now give a definition of probability.

DEFINITION 2

Consider a sample space

$$S = \{s_1, s_2, s_3, \ldots, s_m\}$$

For each k let there be assigned a number w_k to the sample point s_k such that

(1) $w_k \geq 0$

(2) $w_1 + w_2 + w_3 + \cdots + w_m = 1$

The number w_k is called the *weight* of the sample point s_k.

Example 5
If we let

$$w_k = \frac{1}{n(S)} = \frac{1}{m} \quad \text{for} \quad k = 1, 2, 3, \ldots, m$$

then conditions (1) and (2) of the definition are satisfied. Here the sample space consists of equiprobable events.

DEFINITION 3
Let $A \subset S$. Then the sum of weights of the sample points in A is designated $P(A)$ and called the *probability* of the event A.

Example 6
If the members of a sample space S are assigned the weights $1/n(S)$, then for an event $A \subset S$ we have

$$P(A) = \frac{n(A)}{n(S)}$$

Example 7
A coin is tossed six times. What is the probability of exactly 2 Heads?

SOLUTION
Let the sample space be S and let A be the subset of possible outcomes with exactly 2 Heads. From the fundamental counting principle stated in Section 3.5 we know that

$$n(S) = 2^6 = 64$$

and from Example 6 of Section 3.5 we know that

$$n(A) = \binom{6}{4} = 15$$

Hence, the probability of 2 Heads is

$$P(A) = \frac{n(A)}{n(S)} = \frac{15}{64}$$

We conclude this section with a situation in which the occurrence of sample points is not equiprobable.

Example 8

In a contest, a contestant can choose independently from among five questions $q_1, q_2, \ldots, q_5$, devised so that q_1 is considered twice as difficult as q_2, q_2 is considered twice as difficult as q_3, and so on. What weights should be assigned to each question to get a fair probability for answering a given question?

SOLUTION

Let the respective weights be $w_1, w_2, \ldots, w_5$. Then

$$w_2 = 2w_1$$
$$w_3 = 2w_2 = 4w_1$$
$$w_4 = 2w_3 = 8w_1$$
$$w_5 = 2w_4 = 16w_1$$

Since

$$w_1 + w_2 + w_3 + w_4 + w_5 = 1$$

we must have

$$w_1 + 2w_1 + 4w_1 + 8w_1 + 16w_1 = 31w_1 = 1$$

Hence, $w_1 = \frac{1}{31}$, and the respective weights should be

$$w_1 = \frac{1}{31} \qquad w_2 = \frac{2}{31} \qquad w_3 = \frac{4}{31} \qquad w_4 = \frac{8}{31} \qquad w_5 = \frac{16}{31}$$

We note in passing that a general formula for the weights is

$$w_k = \frac{2^{k-1}}{2^5 - 1}$$

EXERCISES

1. A coin is tossed twice. Proceeding as in Examples 2 and 3,

    (a) Find the sample space S.
    (b) What is the probability of 2 Heads?
    (c) What is the probability of 1 Head?
    (d) What is the probability of 1 or more Heads?

2. A coin is tossed four times.

   (a) Find the sample space S.
   (b) What is the probability of 2 Heads?
   (c) What is the probability of 3 Heads?
   (d) What is the probability of no Heads?

3. Is the probability of no Tails with four tosses of a coin greater than the probability of this event with five tosses of a coin?

4. If a coin is tossed m times or $m + 1$ times, when is the probability of 1 Head greater?

5. A sample space S has four elements, $S = \{s_1, s_2, s_3, s_4\}$. A weight w_k is assigned to s_k for $k = 1, 2, 3, 4$. Which of the following assignments determines a probability on the events of S?

   (a) $w_1 = w_2 = w_3 = w_4 = \dfrac{1}{4}$

   (b) $w_1 = w_2 = \dfrac{1}{8}, w_3 = w_4 = \dfrac{3}{8}$

   (c) $w_1 = \dfrac{1}{8}, w_2 = \dfrac{1}{6}, w_3 = \dfrac{1}{4}, w_4 = \dfrac{1}{2}$

   (d) $w_1 = w_2 = \dfrac{1}{8}, w_3 = w_4 = \dfrac{1}{4}$

   (e) $w_1 = 1, w_2 = w_3 = w_4 = 0$

6. Figure 4.3 presents a sample space S schematically. Each of the elements (points) of S has weight $\frac{1}{12}$. Find the following probabilities.

   (a) $P(A)$, $P(B)$, and $P(C)$
   (b) $P(A \cup B)$
   (c) $P(A \cup B \cup C)$

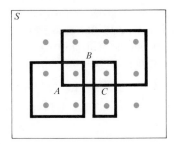

TABLE 4.2. THE SAMPLE SPACE OF A PAIR OF DICE

(1, 1)	(1, 2)	(1, 3)	(1, 4)	(1, 5)	(1, 6)
(2, 1)	(2, 2)	(2, 3)	(2, 4)	(2, 5)	(2, 6)
(3, 1)	(3, 2)	(3, 3)	(3, 4)	(3, 5)	(3, 6)
(4, 1)	(4, 2)	(4, 3)	(4, 4)	(4, 5)	(4, 6)
(5, 1)	(5, 2)	(5, 3)	(5, 4)	(5, 5)	(5, 6)
(6, 1)	(6, 2)	(6, 3)	(6, 4)	(6, 5)	(6, 6)

---

7. A pair of dice is rolled. There are $6 \times 6 = 36$ possible outcomes in the sample space and these are listed in Table 4.2 in the form of *ordered pairs* (x, y). This means that the x entry always belongs to the same die; the y entry, to the other die. If you wish, you can imagine the one die to be yellow, the other red.

   (a) What is the sample space S when we take for the outcome the sum of the dots showing on the two dice?
   (b) If the occurrence of ordered pairs (x, y) is equiprobable, each is assigned the weight 1/36. Using this, find the probability of each sample point in S.

8. In a prize competition, 1,000 contestants compete for one first prize, two second prizes, and ten third prizes. What is a contestant's probability of winning

   (a) First prize?
   (b) A second prize?
   (c) Any prize?

9. If, according to a survey, one person in 2,000 has disease x, what is the probability that a randomly chosen person in a 200 million population has this disease?

10. In a national promotion campaign, a manufacturer offers 3 first prizes, 10 second prizes, 1,000 third prizes, and 5,000 consolation prizes. It is estimated that 12 million people will compete for the prizes. Based on this estimate, what is a participant's probability of winning

    (a) First prize?
    (b) Second prize?
    (c) Third prize?
    (d) Any prize?

11. An urn contains ten black balls and four red balls, all with an equal probability of being drawn (each ball has weight 1/14). If one ball is drawn,

     (a) What is the probability that it is red?
     (b) What is the probability that it is black?
     (c) What is the probability that it is red or black?

12. An urn contains three black and two red balls. The balls are drawn in *pairs*, each pair with an equal probability of being drawn.

     (a) What is the size of the sample space S?
     (b) What is the probability of drawing a red pair?
     (c) What is the probability of drawing a black pair?
     (d) What is the probability of drawing a red-black pair?

13. An urn contains four black and two red balls. Each black ball has weight 1/8 and each red ball has weight 1/4. If one ball is drawn,

     (a) What is the probability that it is red?
     (b) What is the probability that it is black?

14. An urn contains four black, four red, and four green balls, all with an equal probability of being drawn (each ball has weight 1/12). If one ball is drawn,

     (a) What is the probability that it is green?
     (b) What is the probability that it is red or green?
     (c) What is the probability that it is not red?

15. This exercise refers to Example 8.

     (a) What is the probability of answering question q_1 or q_5?
     (b) A contestant will win the same prize if he answers question q_5 or *at least one* of the remaining questions. Which alternative should he pick?

16. In this exercise, S consists of the 52 possible outcomes when a single card is drawn from a regular deck of playing cards. The event that any card x is drawn has probability $\frac{1}{52}$. Find $P(A)$ if

     (a) $A = \{x | x \text{ is an ace}\}$.
     (b) $A = \{x | x \text{ is a heart}\}$.

(c) $A = \{x|x \text{ is not a heart}\}$.
(d) $A = \{x|x \text{ is a queen or a king}\}$.
(e) $A = \{x|x \text{ is a red card}\}$.
(f) $A = \{x|x \text{ is a red card or an ace}\}$.
(g) $A = \{x|x \text{ is not a red ace}\}$.

17. In drawing a card from a marked deck, a red card has probability $3/(4 \times 26)$ and a black card has probability $1/(4 \times 26)$. Find $P(A)$ if

(a) $A = \{x|x \text{ is a heart}\}$.
(b) $A = \{x|x \text{ is a king}\}$.
(c) $A = \{x|x \text{ is not an ace}\}$.
(d) $A = \{x|x \text{ is club or diamond}\}$.
(e) $A = \{x|x \text{ is a black card or an ace}\}$.

4.2 SOME PROPERTIES OF PROBABILITY

Consider a sample space S and an event $A \subset S$. In Section 4.1 we defined the *probability* $P(A)$ to be the sum of weights of the members (sample points) of A. The collection of probabilities $\{P(A)\}$ is a "measure" in S, called a *probability measure*. It enables the comparison of events with one another. A direct consequence of Definitions 2 and 3 of Section 4.1 are the following formulas:

$$P(\emptyset) = 0 \tag{1}$$

$$P(S) = 1 \tag{2}$$

$$0 \leq P(A) \leq 1 \tag{3}$$

Thinking of events as subsets of a sample space makes the discussion of probability conceptually easier because we can use set properties and Venn diagrams. Before introducing an important formula in computations we introduce the concept of exclusive events:

DEFINITION 1

The events A and B are *mutually exclusive* if $A \cap B = \emptyset$.

Thus, exclusive events cannot happen on the same trial, and the occurrence of one excludes the occurrence of the other, as is clear in Figure 4.4.

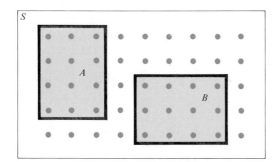

FIGURE 4.4 *Mutually exclusive events* A *and* B

Example 1

Consider the sample space $S = \{1, 2, 3, \ldots, 12\}$ and the events

$A = \{x | x \text{ is divisible by } 2\}$
$B = \{x | x \text{ is divisible by } 3\}$
$C = \{x | x \text{ is divisible by } 7\}$

Which two of these events are mutually exclusive?

SOLUTION

The three events can be expressed as

$A = \{2, 4, 6, 8, 10, 12\}$
$B = \{3, 6, 9, 12\}$
$C = \{7\}$

Hence, A and C, and B and C, are exclusive, but A and B are not exclusive, since $A \cap B = \{6, 12\} \neq \emptyset$.

We can now state the following fact.

THEOREM 1

If A and B are exclusive events, then

$$P(A \cup B) = P(A) + P(B) \tag{4}$$

PROOF

From Formula (1) of Section 3.1 we know that

$$n(A \cup B) = n(A) + n(B)$$

This tells us that the number of sample points in $A \cup B$ equals the number of sample points in A, plus the number of sample points in

B. This tells us, in turn, that each side in (4) has the same sum of weights, and hence equality must hold.

Example 2

A coin is tossed ten times. What is the probability of exactly 4 Heads or exactly 4 Tails?

SOLUTION

Let A be the subset of possible outcomes with 4 Heads, B the subset of possible outcomes with 4 Tails. Since, in this experiment, an outcome with 4 Heads must have 6 Tails, and an outcome with 4 Tails must have 6 Heads, it follows that the events A and B are exclusive, and hence Formula (4) applies. From Section 3.5 we know that if S is the sample space, then

$$n(S) = 2^{10} = 1,024$$

$$P(A) = \frac{\binom{10}{4}}{1,024} = \frac{\binom{10}{6}}{1,024} = P(B)$$

Hence,

$$P(A \cup B) = 2P(B) = 2\frac{\binom{10}{4}}{1,024} = \frac{1}{1,024} \times \frac{2 \times 10!}{4! \, 6!}$$

$$= \frac{1}{1,024} \times \frac{2 \times 10 \times 9 \times 8 \times 7}{4 \times 3 \times 2}$$

$$= \frac{420}{1,024} = \frac{105}{256}$$

An approximate value for the probability of 4 Heads or 4 Tails in ten tosses of a coin is $\frac{2}{5}$.

The next problem will lead us to a more general formula than (4), a formula which holds also when A and B are not exclusive.

Example 3

A coin is tossed four times. What is the probability of 2 or more Heads, or an odd number of Heads?

SOLUTION

Let the sample space be S; let A be the event of 2 or more Heads and B be the event of an odd number of Heads. Then

$$P(A \cup B) = \frac{n(A \cup B)}{n(S)} = \frac{n(A) + n(B) - n(A \cap B)}{n(S)}$$

$$= \frac{n(A)}{n(S)} + \frac{n(B)}{n(S)} - \frac{n(A \cap B)}{n(S)}$$

$$= P(A) + P(B) - P(A \cap B)$$

From Table 4.3 we see that

$$A \cap B = \{\text{HHHT, HHTH, HTHH, THHH}\}$$

and a direct count shows that

$$n(A) = 11$$
$$n(B) = 8$$
$$n(A \cap B) = 4$$
$$n(S) = 2^4 = 16$$

Hence,

$$P(A) = \frac{11}{16}$$

$$P(B) = \frac{8}{16}$$

$$P(A \cap B) = \frac{4}{16}$$

and

$$P(A \cup B) = \frac{11}{16} + \frac{8}{16} - \frac{4}{16} = \frac{15}{16}$$

TABLE 4.3. THE SAMPLE SPACE OF FOUR COIN TOSSES

HHHH	HHHT	HHTT	HTTT	TTTT
	HHTH	HTTH	THTT	
	HTHH	TTHH	TTHT	
	THHH	HTHT	TTTH	
		THHT		
		THTH		

THEOREM 2

If A and B are any events, then

$$P(A \cup B) = P(A) + P(B) - P(A \cap B)$$

The proof of this formula is carried out at the end of this section.

Example 4

In a class of 14 students the following test scores were recorded:

Student	s_1	s_2	s_3	s_4	s_5	s_6	s_7	s_8	s_9	s_{10}	s_{11}	s_{12}	s_{13}	s_{14}
Test a	75	85	40	65	90	75	50	100	70	70	70	80	65	55
Test b	100	85	90	85	95	85	60	95	85	80	50	90	65	70
Average	87.5	85	65	75	92.5	80	55	97.5	77.5	75	60	85	65	62.5

What is the probability that a student will have at least one test score of 90 or higher, or an average of 80 or higher?

SOLUTION

The set of students with at least one test score of 90 or higher is

$$A = \{s_1, s_3, s_5, s_8, s_{12}\}$$

The set of students with an average of 85 or higher is

$$B = \{s_1, s_2, s_5, s_8, s_{12}\}$$

and we see that

$$A \cap B = \{s_1, s_5, s_8, s_{12}\}$$

If S is the sample space, then $n(S) = 14$. Hence, the probability we seek is

$$P(A \cup B) = P(A) + P(B) - P(A \cap B)$$
$$= \frac{n(A)}{n(S)} + \frac{n(B)}{n(S)} - \frac{n(A \cap B)}{n(S)}$$
$$= \frac{5}{14} + \frac{5}{14} - \frac{4}{14} = \frac{3}{7}$$

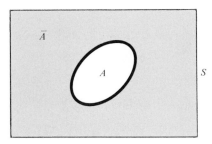

FIGURE 4.5

Complementary events

Consider a sample space S, an event $A \subset S$, and its *complement* $\overline{A}$ (see Figure 4.5). Then

$$A \cup \overline{A} = S \quad \text{and} \quad A \cap \overline{A} = \emptyset$$

Hence, the events A and $\overline{A}$ are exclusive, and Formulas (4) and (2) show that

$$P(A) + P(\overline{A}) = P(S) = 1$$

Solving for $P(A)$ gives

$$P(A) = 1 - P(\overline{A}) \qquad (5)$$

This is an important formula because in practice it is sometimes easier to compute $P(\overline{A})$ than $P(A)$. The events A and $\overline{A}$ are said to be *complementary*. We observe that the probability that an event A will happen or not happen is 1 (it is a "sure thing"). Hence, $P(\overline{A})$ is the probability that A will not happen. We can thus state the following theorem.

THEOREM 3

The probability that an event A will *not* occur is $1 - P(A)$.

The use of this theorem is illustrated in the next example.

Example 5

A shipment of 16 television sets contains 8 defective sets. What is the probability that of 5 television sets picked at random, one or more sets are defective?

SOLUTION

The sample space S in this situation is the set of combinations of 16 objects taken 5 at a time (see Section 3.4). Hence,

$$n(S) = \binom{16}{5} = 4{,}368$$

The events "one or more television sets are defective" and "no television set is defective" are complementary. Thus, if A is the set of combinations of five television sets with one or more defective samples, then $\overline{A}$ is the set of combinations of five television sets with no defective samples, and

$$P(A) = 1 - P(\overline{A}) = 1 - \frac{n(\overline{A})}{n(S)}$$

Since there are eight nondefective television sets,

$$n(\overline{A}) = \binom{8}{5} = 56$$

and hence the answer is

$$P(A) = 1 - \frac{\binom{8}{5}}{\binom{16}{5}} = 1 - \frac{56}{4{,}368}$$

$$= 1 - \frac{7}{546} = \frac{539}{546}$$

An approximate value is $P(A) \approx 987/1{,}000$.

We conclude this section with a proof of Theorem 2.

PROOF OF THEOREM 2
If we put

$$A \cup B = A \cup (B - A)$$

(see Figure 4.6), then

$$A \cap (B - A) = \emptyset$$

and hence by Theorem 1 we have

$$P(A \cup B) = P(A) + P(B - A) \tag{6}$$

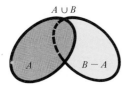

FIGURE 4.6

Since

$$B = (B - A) \cup (A \cap B) \quad \text{and} \quad (B - A) \cap (A \cap B) = \emptyset$$

we again apply Theorem 1. This gives

$$P(B) = P(B - A) + P(A \cap B)$$

or

$$P(B - A) = P(B) - P(A \cap B)$$

Substituting this formula into Formula (6) gives the desired result.

EXERCISES

1. Let $S = \{1, 2, 3, 4\}$, and assign the weight $\frac{1}{10}$ to 1, $\frac{2}{10}$ to 2, $\frac{3}{10}$ to 3, and $\frac{4}{10}$ to 4. Find the following probabilities:

    (a) $P(A \cap B)$ when $A = \{1, 2\}$ and $B = \{2, 3\}$
    (b) $P(A \cup B)$ when $A = \{1, 2\}$ and $B = \{1, 4\}$
    (c) $P(A \cup B)$ when $A = \{1, 3\}$ and $B = \{2, 4\}$

2. Let $S = \{s_1, s_2, s_3, s_4, s_5, s_6\}$; assign the weight $k/21$ to s_k for $k = 1, 2, \ldots, 6$. Find the following probabilities:

    (a) $P(A \cap B)$ when $A = \{s_1, s_2, s_3\}$ and $B = \{s_4, s_5, s_6\}$
    (b) $P(A \cap B)$ when $A = \{s_1, s_3\}$ and $B = \{s_3, s_6\}$
    (c) $P(A \cup B)$ when $A = \{s_2, s_4, s_6\}$ and $B = \{s_1, s_3\}$
    (d) $P(A \cup B)$ when $A = \{s_1, s_2, s_3, s_5\}$ and $B = \{s_1, s_3, s_6\}$

3. Let S be the set of positive integers not exceeding 100; let A be the subset of integers divisible by 5 and B the subset of integers divisible by 10. If $x \in S$, what is the probability that

    (a) $x \in A \cap B$?
    (b) $x \in A \cup B$?

4. In a shipment of 10,000 cartons of eggs, 5 percent contain broken eggs and 8 percent contain stale eggs, but no carton contains both broken and stale eggs. What is the probability that a carton selected at random contains neither broken nor stale eggs?

5. In a shipment of 150 television sets, 12 sets were missing a control knob, 17 sets were misadjusted, and 3 of the sets with a missing control knob were misadjusted. If a set is picked at random, what is the probability that

    (a) It is missing a control knob or it is misadjusted?
    (b) It is only misadjusted?
    (c) It is ready for delivery?

6. As a reward in a maze experiment, a rat could get a food pellet at dispenser A or dispenser B. In an experiment involving 33 rats, 28 rats got a pellet at A and 12 rats got a pellet at B. If a rat is selected at random, what is the probability that it got a pellet at A and at B?

7. Of 100 test papers, 27 contained errors in Part I, 52 contained errors in Part II, and 19 contained errors in both parts. If a test paper is picked at random, what is the probability that

    (a) It is error free?
    (b) It contains errors in both parts?
    (c) It contains errors in Part I or Part II?
    (d) It contains errors in Part II but not in Part I?

8. This exercise refers to Example 4. Find the probabilities of the following events:

    (a) At least one score of 90 or higher *and* an average of 80 or higher.
    (b) At least one score of 90 or higher and an average *below* 80.
    (c) No score over 85, but an average of at least 80.
    (d) Getting a passing grade for the course if each test grade must be at least 65.

9. This exercise refers to Exercise 7 of Section 4.1: A pair of dice is rolled and the sum of the showing dots is taken as the outcome. In this experiment we use the sample space $S = \{2, 3, 4, \ldots, 12\}$. Find the following probabilities:

    (a) $P(2 \text{ or } 3)$

    (b) $P(3 \text{ or } 4 \text{ or } 5)$

    (c) $P(7 \text{ or } 11)$

10.   Referring to Exercise 9, let $x \in S$. Find $P(A \cup B)$ if

    (a) $A = \{x | x \leq 6\}$ $B = \{x | x \geq 6\}$

    (b) $A = \{x | x < 7\}$ $B = \{x | x > 7\}$

    (c) $A = \{x | 4 < x < 8\}$ $B = \{x | 6 < x < 10\}$

    (d) $A = \{x | x \text{ is even}\}$ $B = \{x | x \text{ is divisible by } 4\}$

11.   Find the following probabilities when each of the sample points in Figure 4.7 has equal weight.

    (a) $P(A \cup B)$

    (b) $P(A \cup B \cup C)$

    (c) $P(A - B)$

    (d) $P(A \cap B)$

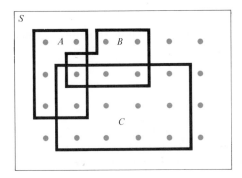

FIGURE 4.7

12.   Test your understanding of the material in Sections 4.1–4.2 with the following True-False questions. Circle your answer.

    (a) $P(A) + P(B) = P(A \cup B) + P(A \cap B)$ T F

    (b) $0 < P(A) < 1$ for any event $A \subset S$ T F

    (c) $P(A \cup B) \leq P(A) + P(B)$ T F

    (d) $P(A - B) = P(A) - P(B)$ T F

    (e)   If $P(A) = 0$, then $A = \emptyset$ T F

    (f)   If $A = \emptyset$, then $P(A) = 0$ T F

    (g)   If $A \subset S$, then $P(A) = \dfrac{n(A)}{n(S)}$ T F

    (h)   For any experiment there is only one possible sample space.   T F

4.3 CONDITIONAL PROBABILITY AND INDEPENDENT EVENTS

In Sections 4.1 and 4.2 we discussed the probabilities of single events. We now consider a more sophisticated type of problem involving sequences of events. In such situations an earlier event may contain information relevant to the occurrence of a later event, and may in fact alter its probability. How this additional information affects probabilities and how it is used is the subject of this section. We begin the discussion with two specific examples.

Example 1
Consider the sample space in Figure 4.8, where each sample point has weight $\frac{1}{20}$. We ask the following two questions about an event x:

(1)   What is the probability that $x \in A$ if $x \in S$?
(2)   What is the probability that $x \in A$ if $x \in B$?

SOLUTION
Question (1) is of the type already studied; namely, since $n(A) = 9$ and $n(S) = 20$, we have

$$P(A) = \frac{n(A)}{n(S)} = \frac{9}{20}$$

To answer question (2) we note that now B plays the role of a sample space, since x is restricted to this set. The part of A which is relevant here is $A \cap B$, and it is seen that the probability we seek is

$$\frac{n(A \cap B)}{n(B)} = \frac{4}{12}$$

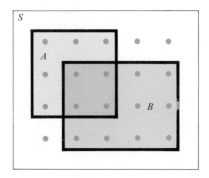

FIGURE 4.8

It comes as no surprise that the probability of the event "$x \in A$" has changed. For this situation we introduce the new symbol

$P(A|B)$, read "the probability of A given B"

and hence

$$P(A|B) = \frac{4}{12}$$

We now observe that

$$\frac{n(A \cap B)}{n(S)} = \frac{\dfrac{n(A \cap B)}{n(S)}}{\dfrac{n(B)}{n(S)}} = \frac{P(A \cap B)}{P(B)} \qquad (7)$$

and thus arrive at the formula

$$P(A|B) = \frac{P(A \cap B)}{P(B)}$$

We shall see in the next example that this formula is not coincidental.

Example 2

In the game in Figure 4.9 you have a set of paths from 0 to the points c_1, c_2, ..., c_7. Each move is equally likely to occur. You can imagine that you enter your piece at 0 and roll a die. Depending on the outcome, you make one of three moves, as indicated in Figure 4.10, and repeat for your next move. The two questions we ask about this game are these:

(1) What is the probability that your piece will reach c_3?
(2) What is the probability that your piece will reach c_3 if it reached a_1?

SOLUTION

(1) The sample space S in this game consists of all paths along straight arrows from 0 to the points c_1, c_2, ..., c_7. From each point, there are three moves, each with three possible outcomes. The total number of paths is therefore $n(S) = 3 \times 3 \times 3 = 27$. The set A of paths from 0 to c_3 is easily found by looking "backward" from c_3 to 0

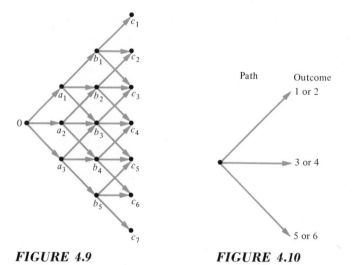

FIGURE 4.9   **FIGURE 4.10**

[see Figure 4.11(a)]. A direct count shows that $n(A) = 6$, and the probability of reaching c_3 is

$$P(A) = \frac{n(A)}{n(S)} = \frac{6}{27}$$

(2)   Once your piece has reached a_1, the move from 0 to a_1 is no longer relevant. The sample space with this information is therefore no longer S, but is the subset B of paths going through a_1 [see Figure 4.11(b)]. Now the game consists of two moves only, each with three

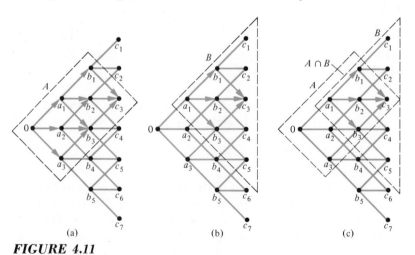

FIGURE 4.11

equiprobable outcomes, giving that $n(B) = 3 \times 3 = 9$. The paths leading from a_1 to c_3 are those contained in the set $A \cap B$ (see Figure 4.11(c)), and clearly $n(A \cap B) = 3$. Now, if x stands for any path from a_1 to c_3, then question (2) can be restated as follows: What is the probability that $x \in A$ if $x \in B$? With the notation $P(A|B)$ introduced in Example 1, that probability is

$$P(A|B) = \frac{n(A \cap B)}{n(B)} = \frac{3}{9} = \frac{1}{3}$$

From (7) we see that once more

$$P(A|B) = \frac{P(A \cap B)}{P(B)}$$

The number $P(A|B)$ is called the *conditional probability* of A, given that B is a certain event. As indicated by the above examples, the following theorem holds.

THEOREM 1

The conditional probability of an event A, given the occurrence of B, is

$$P(A|B) = \frac{P(A \cap B)}{P(B)} \tag{8}$$

Example 3

An urn contains four red and three yellow balls. What is the conditional probability that a pair of balls drawn at random will contain a red ball if it contains a yellow ball?

SOLUTION

To use Formula (8), we shall express this problem in terms of appropriate sets A and B.

Let the red balls be labeled r_1, r_2, r_3, r_4, and the yellow balls y_1, y_2, y_3. Then the sample space S of *all* (unordered) pairs of balls is readily written out (see Figure 4.12). Being interested only in pairs having a red ball, we take for A the set indicated in Figure 4.12. Since one ball is certain to be yellow, we use as sample space the set B. By direct count,

$$n(A \cap B) = 12 \quad \text{and} \quad n(B) = 15$$

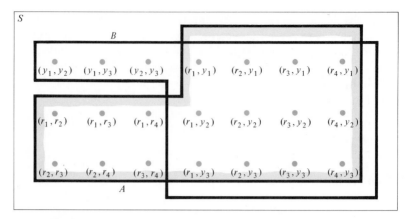

FIGURE 4.12

and hence,

$$P(A|B) = \frac{P(A \cap B)}{P(B)} = \frac{\dfrac{n(A \cap B)}{n(S)}}{\dfrac{n(B)}{n(S)}} = \frac{n(A \cap B)}{n(B)} = \frac{12}{15}$$

An important formula is obtained when Formula (8) is written in the form

$$P(A \cap B) = P(B) \cdot P(A|B) \qquad (9)$$

This formula, called the *product rule*, results when both sides of Formula (8) are multiplied by $P(B)$. It states that the probability that both A and B will occur equals the probability that B will occur, times the conditional probability that A will occur, given that B occurred. Interchanging A and B in Formula (9) gives

$$P(B \cap A) = P(A) \cdot P(B|A)$$

Since $P(B \cap A) = P(A \cap B)$, we arrive at the conclusion that

$$P(A \cap B) = P(A) \cdot P(B|A) = P(B) \cdot P(A|B) \qquad (10)$$

Example 4

Of 100 dozen egg cartons, 20 percent contain stale eggs, and 5 percent of those contain broken eggs. What is the probability that an egg carton, picked at random, will contain both stale and broken eggs?

SOLUTION

Let us use Formula (9). If S is the sample space of egg cartons, B the set of cartons with stale eggs, and A the set of cartons with broken eggs, then

$$n(S) = 100, \quad n(B) = 20, \quad \text{and} \quad n(A \cap B) = 1$$

Hence,

$$P(B) = \frac{n(B)}{n(S)} = \frac{20}{100} = \frac{1}{5} \quad \text{and} \quad P(A|B) = \frac{1}{20}$$

and so

$$P(A \cap B) = \frac{1}{5} \times \frac{1}{20} = \frac{1}{100}$$

Example 5

In a ten-question True-False test it is known that three consecutive answers are true. What is the conditional probability of guessing correctly all answers if it is known which consecutive three answers are true?

SOLUTION

Let S be the set of all possible answers. Consult Figure 4.13 in which three particular questions are answered true, leaving seven questions to be answered. Let the set of possible answers to these questions be B, and let the set of correct answers be A. Then

$$n(S) = 2^{10}, \qquad n(B) = 2^7, \quad \text{and} \quad n(A) = n(A \cap B) = 1$$

Hence,

$$P(B) = \frac{n(B)}{n(S)} = \frac{2^7}{2^{10}} = \frac{1}{2^3} \quad \text{and} \quad P(A \cap B) = \frac{n(A \cap B)}{n(S)} = \frac{1}{2^{10}}$$

and so

$$P(A|B) = \frac{\dfrac{1}{2^{10}}}{\dfrac{1}{2^3}} = \frac{1}{2^7} = \frac{1}{128}$$

Answer	?	?	T	T	T	?	?	?	?	?
Question	1	2	3	4	5	6	7	8	9	10

FIGURE 4.13

Independent events

In certain situations Formula (9) can be replaced by the formula $P(A \cap B) = P(A) \cdot P(B)$. In view of Formula (10), this means that the conditional probability of A equals the probability of A, and likewise for B. This in turn means that one event has no effect on the occurrence of the other, and we say that the events are *independent*. Before giving a formal definition, we illustrate this concept in the following example.

Example 6

A coin is tossed twice. What is the conditional probability for 2 Heads if it is known that the coin shows the same face in the two turns?

SOLUTION

Let us look at the sample space S of this experiment:

$$S = \{HH, HT, TH, TT\}$$

If we are to have two Heads, we must have Head in the first toss. We thus consider the event

$$A = \{HH, HT\}$$

Since it is known that the same face shows, we also consider the set

$$B = \{HH, TT\}$$

We now see that the problem asks for the probability $P(A|B)$. Now, $A \cap B = \{HH\}$, $n(A \cap B) = 1$, and $n(B) = 2$, and hence

$$P(A|B) = \frac{n(A \cap B)}{n(B)} = \frac{1}{2}$$

This example leads to a new concept as follows: We note that

$$P(B) = \frac{n(B)}{n(S)} = \frac{2}{4} = \frac{1}{2}$$

and by Formula (9),

$$P(A \cap B) = P(B) \cdot P(A|B) = \frac{1}{2} \times \frac{1}{2} = \frac{1}{4}$$

Since $P(A) = \frac{1}{2}$, we also have in this case

$$P(A \cap B) = P(A) \cdot P(B)$$

and when this formula holds, we call the events A and B independent.

DEFINITION 1

The events A and B are *independent* if

$$P(A \cap B) = P(A) \cdot P(B) \tag{11}$$

and *dependent* otherwise.

In practice, it may not always be readily decidable if two events are independent, since you may not be able to compute one of the probabilities in Formula (11).

Example 7

A pair of dice is rolled and the sum of the showing dots is added up. Consider the events

$$A = \{9, 10, 11, 12\} \qquad \text{(the sum of the dots is at least 9)}$$
$$B = \{8, 10, 12\} \qquad \text{(the sum of the dots is 8, 10, or 12)}$$

Are these events independent?

SOLUTION

To answer this question, we have to decide if Formula (11) holds. Consulting Exercise 7 in Section 4.1 and Exercise 9 in Section 4.2 we find

$$P(A) = \frac{4}{36} + \frac{3}{36} + \frac{2}{36} + \frac{1}{36} = \frac{10}{36}$$

$$P(B) = \frac{5}{36} + \frac{3}{36} + \frac{1}{36} = \frac{9}{36}$$

and hence

$$P(A) \cdot P(B) = \frac{10}{36} \times \frac{9}{36} = \frac{5}{72}$$

Since

$$A \cap B = \{10, 12\}$$

we have

$$P(A \cap B) = \frac{3}{36} + \frac{1}{36} = \frac{4}{36} = \frac{8}{72}$$

Hence,

$$P(A \cap B) \neq P(A) \cdot P(B)$$

and this tells us that the events A and B are *dependent*.

EXERCISES

1. Referring to Example 1:

    (a) What is the conditional probability $P(B|A)$?
    (b) What is the conditional probability $P(A|\overline{B})$? (Recall that $\overline{B}$ is the complement of B.)
    (c) Are the events $A - B$ and $B - A$ independent?

2. A rat is introduced into the maze in Figure 4.14 at 0. It is equally likely to select any one of the three paths at 0, g_1, g_2, and g_3.

    (a) Find the probability for reaching each terminal point $h_1, h_2, \ldots, h_5$.
    (b) What is the conditional probability of reaching h_4 if the rat reached g_2?
    (c) What is the conditional probability of reaching h_4 if the rat reached g_3?
    (d) What is the conditional probability of reaching h_1 if the rat reached g_3?

3. Referring to Example 2:

    (a) Which of the points $c_1, c_2, \ldots, c_7$ is your piece most likely to reach? Least likely to reach?
    (b) What is the probability that your piece will reach c_3 if it reached a_3?
    (c) Which of the points $c_1, c_2, \ldots, c_7$ is your piece most likely to reach if it reached a_2?

4. In the game in Figure 4.15 you enter your piece at 0, and move it from left to right with the objective of reaching one of the points c_k, $k = 1, 2, 3, \ldots, 11$. The possible moves from each junction are equiprobable, and they are decided with a die.

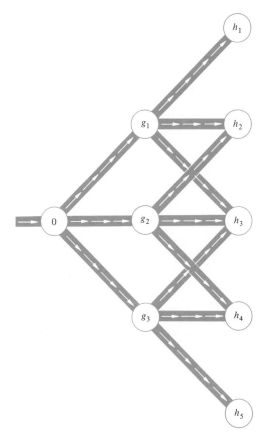

FIGURE 4.14

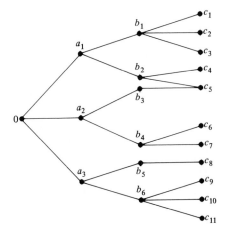

FIGURE 4.15

(a) What is the probability of reaching each of the following points: c_1, c_4, c_5, c_8, c_{11}?
(b) What is the conditional probability of reaching c_5 if a_1 was reached?
(c) What is the conditional probability of reaching c_8 if a_3 was reached?
(d) What is the conditional probability of reaching c_1 or c_4 if a_1 was reached?

5. An urn contains four red and four yellow balls. All things being equal, what is the conditional probability that a pair of balls drawn at random will be yellow if it contains one yellow ball?
6. A pair of cards is drawn from a deck of 52 playing cards. What is the conditional probability for a pair of aces if one ace is included in the pair?
7. To discourage guessing, an instructor phrases the *same* question in four different ways on a ten-question True-False test.

(a) What is the probability of guessing correctly all answers?
(b) What is the conditional probability of guessing correctly all answers if the four questions with the same answer are identified?

8. A deck of 52 playing cards is divided into two by color, and a black and a red card are drawn at random.

(a) What is the probability that both cards are aces?
(b) What is the conditional probability that the red card is an ace, given that the black card is an ace?
(c) Let R be the event of drawing a red ace, B the event of drawing a black ace. Are the events R and B independent?

9. A coin is tossed three times. Which of the following pairs of events are independent?

(a) $A = \{HHH, TTT\}$ $B = \{HHT, HTH, THH\}$
(b) $A = \{HHH, HHT, HTH, THH\}$
      $B = \{HTT, THT, TTH, TTT\}$
(c) $A = \{HHH, TTT\}$ $B = \{TTT, TTH, THT, HTT\}$

10. A pair of dice is rolled (see Example 7). Which of the following pairs of events are independent?

(a) $A = \{9, 10, 11\}$ $B = \{4, 7, 10\}$
(b) $A = \{9, 11\}$ $B = \{7, 10, 11\}$
(c) $A = \{2, 3, 4\}$ $B = \{2, 10, 11\}$
(d) $A = \{2, 3, 4, 5, 6\}$ $B = \{2, 3, 4, 5, 6\}$

11. Let $A_1, A_2, \ldots, A_r$ be exclusive events. Show that the following generalization of Formula (11) holds:

$$P(A_1 \cap A_2 \cap \cdots \cap A_r) = P(A_1) \cdot P(A_2) \cdot \cdots \cdot P(A_r)$$

Hint:

Put $B_1 = A_2 \cap A_3 \cap \cdots \cap A_r$ and apply Formula (11) to $A_1 \cap B_1$. Next, put $B_2 = A_3 \cap A_4 \cap \cdots \cap A_r$ and apply Formula (11) to $A_2 \cap B_2$, and so on.

4.4 BAYES' FORMULA

With the reasoning used in Section 4.3, we shall develop here a formula for solving a class of problems of the following variety.

Example 1

An automobile assembly plant receives engines from manufacturing plants a_1, and a_2, referred to here as a_1-engines and a_2-engines respectively. From past experience it is estimated that 1 percent of the a_1-engines are defective and that 3 percent of the a_2-engines are defective. Suppose 900 a_1-engines and 1700 a_2-engines are on hand, and an engine picked at random is found to be defective. Based on the above estimate, what is the probability that it is an a_1-engine?

SOLUTION

If we designate the set of a_1-engines by A_1, the set a_2-engines by A_2, then the sample space for this problem is $S = A_1 \cup A_2$. If we let B stand for the set of defective engines in S (see Figure 4.16), then the problem asks for the conditional probability $P(A_1|B)$.

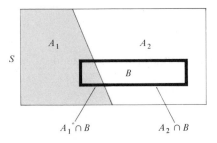

FIGURE 4.16

From the data of this problem we have the following information:

$$P(B|A_1) = \frac{1}{100} \qquad \text{(1 percent)}$$

$$P(B|A_2) = \frac{3}{100} \qquad \text{(3 percent)}$$

$$P(A_1) = \frac{900}{900 + 1{,}700} = \frac{9}{26} \tag{12}$$

$$P(A_2) = \frac{1{,}700}{900 + 1{,}700} = \frac{17}{26}$$

How can we use this information directly to solve this problem? Let us examine the formula

$$P(A_1|B) = \frac{P(A_1 \cap B)}{P(B)} \tag{13}$$

We recall from Section 4.3 [Formula (10)] that

$$P(A_1 \cap B) = P(A_1)P(B|A_1) \tag{14}$$

Next, we observe that $[A_1 \cap B, \ A_2 \cap B]$ is a dichotomy of B (see Figure 4.16). This tells us that

$$B = (A_1 \cap B) \cup (A_2 \cap B)$$

where the sets $A_1 \cap B$ and $A_2 \cap B$ are mutually exclusive; by Theorem 1 in Section 4.2, therefore,

$$\begin{aligned} P(B) &= P(A_1 \cap B) + P(A_2 \cap B) \\ &= P(A_1)P(B|A_1) + P(A_2)P(B|A_2) \end{aligned} \tag{15}$$

Substituting (14) and (15) into (13) gives

$$P(A_1|B) = \frac{P(A_1)P(B|A_1)}{P(A_1)P(B|A_1) + P(A_2)P(B|A_2)} \tag{16}$$

We are thus able to express $P(A_1|B)$ directly in terms of the data in (12), and we have

$$P(A_1|B) = \frac{\dfrac{9}{26} \times \dfrac{1}{100}}{\dfrac{9}{26} \times \dfrac{1}{100} + \dfrac{17}{26} \times \dfrac{3}{100}} = \frac{9}{9 + 51} = \frac{9}{60} = \frac{3}{20}$$

Formula (16) is known as *Bayes' formula* for a dichotomy of the sample space. For a general partition of the sample space, the following is true.

Bayes' formula

Let $[A_1, A_2, \ldots, A_r]$ be a partition of a sample space S, and let B be a given event in S. Then, for $k = 1, 2, 3, \ldots, r,$

$$P(A_k|B)$$
$$= \frac{P(A_k)P(B|A_k)}{P(A_1)P(B|A_1) + P(A_2)P(B|A_2) + \cdots + P(A_r)P(B|A_r)} \quad (17)$$

Example 2

Table 4.4 gives the responses of 400 students polled about their major and future plans. A student picked at random from this group turned out to be undecided about his career. What is the probability that the student is an undeclared major?

TABLE 4.4

Major	Number of students	Undecided about future career, %
Life sciences	70	50
Mathematics	30	20
Engineering	20	20
Humanities	100	21
Undeclared	180	70

SOLUTION

We shall solve this problem with the help of Formula (17). To do this, the problem must be expressed in terms of suitable sets A_1, $A_2, \ldots, A_r$ and B. Thus, consider the following partition of the sample space of students polled:

$A_1 = \{\text{life sciences majors}\}$
$A_2 = \{\text{mathematics majors}\}$
$A_3 = \{\text{engineering majors}\}$
$A_4 = \{\text{humanities majors}\}$
$A_5 = \{\text{undeclared majors}\}$

Let B be the set of students in the poll which were undecided about their careers. The problem then asks for the conditional probability $P(A_5|B)$. Hence, we use Formula (17) with $k = 5$ and $r = 5$. From the data in Table 4.4 we have the following:

$$P(A_1) = \frac{7}{40} \qquad P(B|A_1) = \frac{50}{100}$$

$$P(A_2) = \frac{3}{40} \qquad P(B|A_2) = \frac{20}{100}$$

$$P(A_3) = \frac{2}{40} \qquad P(B|A_3) = \frac{20}{100}$$

$$P(A_4) = \frac{10}{40} \qquad P(B|A_4) = \frac{21}{100}$$

$$P(A_5) = \frac{18}{40} \qquad P(B|A_5) = \frac{70}{100}$$

Hence,

$$P(A_5)P(B|A_5) = \frac{18}{40} \times \frac{70}{100} = \frac{1,260}{4,000}$$

and

$$P(A_1)P(B|A_1) + \cdots + P(A_5)P(B|A_5)$$
$$= \frac{7}{40} \times \frac{50}{100} + \frac{3}{40} \times \frac{20}{100} + \frac{2}{40} \times \frac{20}{100} + \frac{10}{40} \times \frac{21}{100} + \frac{18}{40} \times \frac{70}{100}$$
$$= \frac{1}{4,000} (350 + 60 + 40 + 210 + 1260) = \frac{1,920}{4,000}$$

giving as final result

$$P(A_5|B) = \frac{1,260}{1,920} = \frac{21}{32}$$

Example 3
Referring to Example 2, what is the probability that the student is a mathematics *or* engineering major?

SOLUTION
Now we are asked for the conditional probability $P(A_2 \cup A_3|B)$. We thus take $[A_1, A_2 \cup A_3, A_4, A_5]$ as the partition of the sample space and we use the formula

$P(A_2 \cup A_3|B)$

$$= \frac{P(A_2 \cup A_3)P(B|A_2 \cup A_3)}{P(A_1)P(B|A_1) + P(A_2 \cup A_3)P(B|A_2 \cup A_3) + P(A_4)P(B|A_4) + P(A_5)P(B|A_5)}$$

Since A_2 and A_3 are disjoint, we have

$$P(A_2 \cup A_3) = P(A_2) + P(A_3) = \frac{3}{40} + \frac{2}{40} = \frac{5}{40}$$

We see that 20 percent of the students in $A_2 \cup A_3$ are undecided about careers, and so

$$P(B|A_2 \cup A_3) = \frac{20}{100}$$

Hence,

$$P(A_2 \cup A_3)P(B|A_2 \cup A_3) = \frac{5}{40} \times \frac{20}{100} = \frac{100}{4{,}000}$$

and

$$P(A_1)P(B|A_1) + P(A_2 \cup A_3)P(B|A_2 \cup A_3)$$
$$+ P(A_4)P(B|A_4) + P(A_5)PB|A_5)$$
$$= \frac{7}{40} \times \frac{50}{100} + \frac{5}{40} \times \frac{20}{100} + \frac{10}{40} \times \frac{21}{100} + \frac{18}{40} \times \frac{70}{100}$$
$$= \frac{1}{4{,}000} (350 + 100 + 210 + 1260) = \frac{1{,}920}{4{,}000}$$

This gives

$$P(A_2 \cup A_3|B) = \frac{100}{1{,}920} = \frac{5}{96}$$

Example 4

Consider the diagram in Figure 4.17. Let the sample space S consist of all paths from 0 to one of the points $c_1, c_2, \ldots, c_{11}$; a direct count gives $n(S) = 17$. Consider the partition $[A_1, A_2, A_3]$ of S with cells $A_k = \{$paths through $a_k\}$, $k = 1, 2, 3$. Let $B = \{$paths through $b_2\}$. Find $P(A_1|B)$.

SOLUTION

The following information is available directly from Figure 4.17:

$$n(A_1) = 7 \qquad P(A_1) = \frac{7}{17} \qquad P(B|A_1) = \frac{4}{7}$$

$$n(A_2) = 6 \qquad P(A_2) = \frac{6}{17} \qquad P(B|A_2) = \frac{4}{6}$$

$$n(A_3) = 4 \qquad P(A_3) = \frac{4}{17} \qquad P(B|A_3) = 0$$

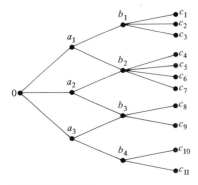

FIGURE 4.17

Hence,

$$P(A_1)P(B|A_1) = \frac{7}{17} \times \frac{4}{7} = \frac{4}{17}$$

$$P(A_2)P(B|A_2) = \frac{6}{17} \times \frac{4}{6} = \frac{4}{17}$$

$$P(A_3)P(B|A_3) = 0$$

and Formula (17) gives

$$P(A_1|B) = \frac{\frac{4}{17}}{\frac{4}{17} + \frac{4}{17} + 0} = \frac{1}{2}$$

EXERCISES

1. Referring to Example 1: What is the probability that the engine picked is an a_2-engine?

2. Referring to Example 2: Find the probability that the student is

    (a) A humanities major.
    (b) A life sciences major.
    (c) Not an undeclared major.

3. This exercise refers to Example 4 and Figure 4.17.

    (a) Find $P(A_2|B)$.
    (b) Find $P(A_1 \cup A_2|B)$.

4. This exercise refers to Example 4 and Figure 4.17. Let

    $B_1 = \{\text{paths through } b_1 \text{ or } b_2\}$
    $B_2 = \{\text{paths through } b_2 \text{ or } b_3 \text{ or } b_4\}$

Find

(a) $P(A_1|B_1)$
(b) $P(A_2|B_1)$
(c) $P(A_1 \cup A_2|B_1)$
(d) $P(A_1|B_2)$
(e) $P(A_3|B_2)$

5. Referring to Figure 4.18: Consider the sample space S of all paths from 0 to the points c_1, c_2, ..., c_{14}. Let $[A_1, A_2, A_3]$ be a partition of S with cells $A_k = \{\text{paths through } a_k\}$, $k = 1, 2, 3$. Let

$B = \{\text{paths through } b_3\}$
$C = \{\text{paths through } b_2 \text{ or } b_4 \text{ or } b_6 \text{ or } b_8\}$
$D = \{\text{paths } not \text{ through } b_1 \text{ or } b_2\}$

Find

(a) $P(A_1|B)$
(b) $P(A_1|C)$
(c) $P(A_1 \cup A_2|B)$
(d) $P(A_2|C)$
(e) $P(A_2|D)$
(f) $P(A_3|C)$
(g) $P(A_3|D)$
(h) $P(A_2 \cup A_3|C)$
(i) $P(A_2 \cup A_3|D)$

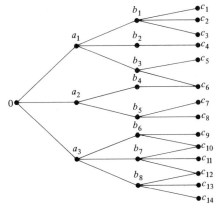

FIGURE 4.18

6. An urn contains twelve red and six yellow balls. A ball is drawn; if it is yellow, then the experiment ends. If it is red, then a second ball is drawn.

   (a) What is the probability of drawing the second ball?
   (b) If a second ball is drawn, what is the probability that it is yellow?
   (c) What is the probability of drawing two red balls?

7. A supply of ball bearings is estimated to have the following sources and distribution.

BALL BEARINGS (IN PERCENT)

Source	Acceptable	Defective	Total
Machine A	35	5	40
Machine B	23	2	25
Machine C	34	1	35

A sample picked at random is found to be defective. Based on the above estimate,

   (a) What is the probability that it was manufactured by machine A?
   (b) What is the probability that it was manufactured by machine C?
   (c) What is the probability that it was not manufactured by machine C?

8. Suppose that in a student population containing 60 percent females, 12 percent of the males and 7 percent of the females are left-handed. A student picked at random is found to be left-handed. What is the probability that

   (a) The student is female?
   (b) The student is male?

9. Referring to Exercise 8, what is the probability that a student picked at random is female if right-handed?

10. A proposition was submitted to the voters in a state election. The voter's party affiliation and vote on the proposition was as follows.

Party	Registration in Percent	Percent Who Voted in Favor of the Proposition
Democratic	40	80
Republican	30	10
Independent	20	90
Others	10	40

A voter is picked at random.

(a) What is the probability that he voted in favor of the proposition?
(b) If he voted in favor of the proposition, what is the probability that he is a Democrat?
(c) If he voted against the proposition, what is the probability that he is an Independent?
(d) If he voted against the proposition, what is the probability that he is *not* a Republican?

4.5 REPEATED EVENTS WITH TWO POSSIBLE OUTCOMES

Many experiments consist of repeating a single event. When the probability of an event remains the same under repetitions, these may be regarded as independent events in the sense of Definition 1 of Section 4.3. A case in point is the repeated tossing of a single coin. Thus, if E_k is a possible outcome of an event on a kth repetition, and the event is repeated m times, then

$$P(E_1 \cap E_2 \cap \cdots \cap E_m) = P(E_1) \cdot P(E_2) \cdot \cdots \cdot P(E_m) \quad (18)$$

(See Exercise 11 in Section 4.3.) We illustrate the consequences of this assumption with an example.

Example 1
A disk is divided into three equal sectors, one colored red and two colored green (see Figure 4.19), and a dial is spun ten times. All things being equal, what is the probability of green coming up exactly six times?

FIGURE 4.19

SOLUTION

Each trial in this game has two possible outcomes: "green" and "red," with respective probabilities $\frac{2}{3}$ and $\frac{1}{3}$. We note that the events "green" and "red" are *complementary* (see Section 4.2).

Consider any possible outcome of ten spins of the dial in which "green" came up six times and "red" came up four times. If in Formula (18) we let $E_1, E_2, \ldots, E_6$ stand for "green," and we let $E_7, E_8, E_9,$ and E_{10} stand for "red," then we can use it here with $m = 10$. The probability of such a particular outcome is seen to be

$$\underbrace{\left(\frac{2}{3} \times \frac{2}{3} \times \cdots \times \frac{2}{3}\right)}_{6 \text{ times}} \times \underbrace{\left(\frac{1}{3} \times \cdots \times \frac{1}{3}\right)}_{4 \text{ times}} = \left(\frac{2}{3}\right)^6 \left(\frac{1}{3}\right)^4$$

According to Section 3.6, however, there are $\binom{10}{6}$ possible outcomes. This is so because we are dealing here with the number of combinations of ten elements taken six at a time. Hence, the probability of getting "green" exactly six times is

$$\binom{10}{6}\left(\frac{2}{3}\right)^6\left(\frac{1}{3}\right)^4$$

This answer can be left in this form. It can be computed with a desk calculator or a table, and we find it to be

$$\frac{10!}{6!4!} \times \frac{2^6}{3^6} \times \frac{1}{3^4} = \frac{10 \times 9 \times 8 \times 7}{4 \times 3 \times 2} \times \frac{64}{59,049} \approx 0.23$$

We now make the following observations:

$\frac{2}{3}$ is the probability of the event "green" occurring

$\frac{1}{3} = 1 - \frac{2}{3}$ The probability of the event *not* occurring

$\binom{10}{6}$ is the binomial coefficient of the term $(\frac{2}{3})^6(\frac{1}{3})^4$ in the binomial expansion of $(\frac{2}{3} + \frac{1}{3})^{10} = 1$

Here, then, is one more instance in which the remarkable numbers $\binom{m}{k}$ appear. A generalization of the reasoning used in Example 1 yields the following theorem.

The Binomial Rule

Consider a two-outcome experiment involving an event E and its complement $\overline{E}$. In each of m repetitions, let p be the probability of E occurring. Then the probability that E will occur exactly k times is

$$\binom{m}{k}p^k(1-p)^{m-k} \tag{19}$$

Probabilities involving Formula (19) are called *binomial probabilities*. To illustrate the use of the formula in connection with repeated events, we consider the following example.

Example 2
Referring to Example 1, what is the probability of coming up with "green" at least twice with ten spins of the dial?

SOLUTION
For "green" to come up exactly k times in this game, the probability is

$$\binom{10}{k}\left(\frac{2}{3}\right)^k\left(\frac{1}{3}\right)^{10-k}$$

To get "green" at least twice means to get "green" exactly twice, or exactly three times, or exactly four times, and so on. All these events are exclusive (see Section 4.2), and hence the answer to the problem is

$$P = \binom{10}{2}\left(\frac{2}{3}\right)^2\left(\frac{1}{3}\right)^8 + \binom{10}{3}\left(\frac{2}{3}\right)^3\left(\frac{1}{3}\right)^7 + \binom{10}{4}\left(\frac{2}{3}\right)^4\left(\frac{1}{3}\right)^6$$
$$+ \cdots + \binom{10}{10}\left(\frac{2}{3}\right)^{10}$$

By Formula (2) of Section 3.5, however,

$$1 = \left(\frac{2}{3}+\frac{1}{3}\right)^{10} = \binom{10}{0}\left(\frac{1}{3}\right)^{10} + \binom{10}{1}\left(\frac{2}{3}\right)\left(\frac{1}{3}\right)^9$$
$$+ \underbrace{\binom{10}{2}\left(\frac{2}{3}\right)^2\left(\frac{1}{3}\right)^8 + \cdots + \binom{10}{10}\left(\frac{2}{3}\right)^{10}}_{P}$$

Solving for P gives

$$P = 1 - \binom{10}{0}\left(\frac{1}{3}\right)^{10} - \binom{10}{1}\left(\frac{2}{3}\right)\left(\frac{1}{3}\right)^{9}$$

$$= 1 - \left(\frac{1}{3}\right)^{10} - 10 \times 2 \times \left(\frac{1}{3}\right)^{10} = 1 - 21\left(\frac{1}{3}\right)^{10}$$

Using a desk calculator or a table, we find that

$$1 - 21\left(\frac{1}{3}\right)^{10} = 1 - \frac{21}{59,049} = 1 - 0.00036 = 0.99964$$

Thus, the probability is very high.

Example 3
Under normal conditions, a machine manufacturing magnetic recording tape turns out 99 percent good tape. If ten reels are selected at random, what is the probability that two contain bad tape?

SOLUTION
We assume that the probability for each defective reel of tape is $\frac{1}{100}$. We thus regard the production of bad tape as repeated independent events, and so the binomial rule applies. We use Formula (19) with $m = 10$, $k = 2$, and $p = \frac{1}{100}$. This gives the probability

$$\binom{10}{2}\left(\frac{1}{100}\right)^{2}\left(\frac{99}{100}\right)^{8}$$

By the binomial formula (Section 3.5) we have the approximation

$$\left(\frac{99}{100}\right)^{8} = \left(1 - \frac{1}{100}\right)^{8} \approx 1 - \frac{8}{100} = \frac{98}{100}$$

We can thus calculate the following approximate value:

$$\binom{10}{2}\left(\frac{1}{100}\right)^{2}\left(\frac{99}{100}\right)^{8} \approx \frac{10 \times 9}{2} \times \frac{1}{10,000} \times \frac{98}{100}$$

$$= \frac{441}{100,000} \approx \frac{44}{10,000} = \frac{11}{2,500}$$

Example 4
A family has n children, $n \geq 3$. Assuming the probability of birth of both sexes to be equiprobable, what is the probability that exactly three children are girls?

SOLUTION

We use Formula (19) with $m = n$, $k = 3$, and $p = \frac{1}{2}$. This gives for the probability of exactly three girls among n children:

$$\binom{n}{3}\left(\frac{1}{2}\right)^{3}\left(\frac{1}{2}\right)^{n-3} = \binom{n}{3}\left(\frac{1}{2}\right)^{n} = \binom{n}{3}\frac{1}{2^{n}}$$

Table 4.5 lists the probabilities of exactly three girls when the number of children is $n = 3, 4, 5, \ldots, 10$, and these values are plotted in Figure 4.20. The reader may be surprised to discover that the probability has largest value among five and six children.

In formulating the binomial rule, it was stipulated that the experiment must have only two outcomes. Under the right circum-

TABLE 4.5.   THE PROBABILITY OF 3 GIRLS OUT OF n CHILDREN

No. of children n	Probability
3	$\frac{16}{128}$
4	$\frac{32}{128}$
5	$\frac{40}{128}$
6	$\frac{40}{128}$
7	$\frac{35}{128}$
8	$\frac{28}{128}$
9	$\frac{21}{128}$
10	$\frac{15}{128}$

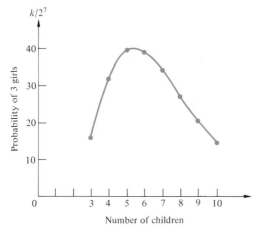

FIGURE 4.20

stances, however, this rule can be applied to multiple-outcome experiments. Such a situation is considered next.

Example 5
A single die is rolled five times. What is the probability of exactly four 6's?

SOLUTION

Although there are six possible outcomes with each throw of a die, we can regard here the outcomes as only two: 6 and not-6. Since 6 has probability $\frac{1}{6}$, and not-6 has probability $\frac{5}{6}$, the binomial rule gives

$$\binom{5}{4}\left(\frac{1}{6}\right)^4\left(\frac{5}{6}\right) = \frac{25}{6^5} = \frac{25}{7,776} \approx \frac{25}{7,775} = \frac{1}{311}$$

Sometimes it is difficult to decide what formula or reasoning to apply to a given problem. Experiments may seem similar when they are not, or dissimilar when they are not. In the following problem we illustrate this with a card-drawing experiment.

Example 6
Consider an ordinary deck of 52 playing cards. What is the probability of drawing three kings and one queen if

(1)   Four cards are drawn together.
(2)   Four cards are drawn *one at a time without replacing* a drawn card in the deck.
(3)   Four cards are drawn *one at a time and replaced* in the deck before the next card is drawn.

SOLUTION

(1)   This experiment involves no repetition and it is of the type discussed in Section 4.2. The sample space consists of the combinations of 52 cards taken four at a time, and hence $n(S) = \binom{52}{4}$ (see Section 3.6). The number of ways three kings can be picked out of four is $\binom{4}{3}$, and since the fourth card can be any queen, there are $\binom{4}{1}\binom{4}{3}$ favorable outcomes. Hence, the probability is

$$\frac{\binom{4}{1}\binom{4}{3}}{\binom{52}{4}} = \frac{4 \times 4}{\dfrac{52 \times 51 \times 50 \times 49}{4 \times 3 \times 2}} = \frac{16}{270,725}$$

(2)   The number of ways in which four cards can be drawn without replacement is $52 \times 51 \times 50 \times 49$, and hence $n(S) = 52 \times 51 \times 50 \times 49$. To get the number of ways in which three kings and one queen can be drawn, we must multiply the number $\binom{4}{1}\binom{4}{3}$ by 4!, the number of permutations on the four cards, since order is relevant here. Hence, the probability is

$$\frac{\binom{4}{1}\binom{4}{3}4!}{52 \times 51 \times 50 \times 49} = \frac{\binom{4}{1}\binom{4}{3}}{\dfrac{52 \times 51 \times 50 \times 49}{4!}} = \frac{\binom{4}{1}\binom{4}{3}}{\binom{52}{4}} = \frac{16}{270,725}$$

We observe that experiments (1) and (2) have the same probability. Thinking about it, we see intuitively why this should be so, because in each of these two experiments we are ultimately dealing with combinations.

(3)   To this problem we can apply Formula (19). Whether the event is one king or one queen, the probability is $p = \frac{4}{52}$. The experiment is repeated $m = 4$ times, and since we want the event to occur $k = 4$ times, we have for the probability

$$\binom{4}{4}\left(\frac{4}{52}\right)^4\left(\frac{48}{52}\right)^0 = \left(\frac{1}{13}\right)^4 = \frac{1}{28,561}$$

EXERCISES

1.   Referring to Example 1: Let the dial be spun seven times. What is the probability of

   (a)   *Green* occurring every time?
   (b)   *Red* occurring exactly six times?
   (c)   *Red* occurring at least six times?

2.   Suppose that the game in Figure 4.19 is so biased that green and red are equiprobable. What is the probability of *green* occurring exactly eight times when the dial is spun ten times?

3.   Referring to Example 3: Suppose that the machine turns out only 95 percent good tape. If five reels are picked at random, what is the probability that one of these contains bad tape?

4.   A record-player manufacturer uses motors including an estimated 1 percent defective samples. Assuming all defective units to be equiprobable, what is the probability that a random selection of five units will contain three defective ones?

5. Referring to Example 4: Consider families with 3, 4, 5, ..., 10 children, respectively.

    (a) Which families are *most likely* to have exactly three girls?
    (b) Which families are *least likely* to have exactly three girls?

6. A family has five children. Assuming the birth of both sexes to be equiprobable, what is the probability of

    (a) At least one boy?
    (b) At most one boy?
    (c) Exactly two boys?
    (d) Exactly three boys?

7. A single die is rolled five times. What is the probability of

    (a) All 3's?
    (b) Exactly four 3's?
    (c) Exactly one 3?

8. A pair of dice is rolled five times and the showing dots are added up. What are the probabilities of the following outcomes:

    (a) Five 7's?
    (b) Five 2's?
    (c) Three 7's?
    (d) Three 2's?
    (e) No 4?
    (f) No 12?

9. What is the probability of $m/2$ Heads when a coin is tossed m times for $m = 4, 8, 12, 16, 20$?

10. Suppose that in a population the birth of both sexes is equiprobable. Of m births, what is the probability for $(m/2) - 1$ girls when $m = 2, 4, 6, 8, 10$?

11. A set of dominoes consists of 28 pieces, as shown in Figure 4.21. Five pieces are picked at random, one at a time, replacing each piece before the next one is picked. Find the probabilities for the following events:

    (a) A double 6 each time
    (b) A double 6 exactly three times
    (c) A double each time

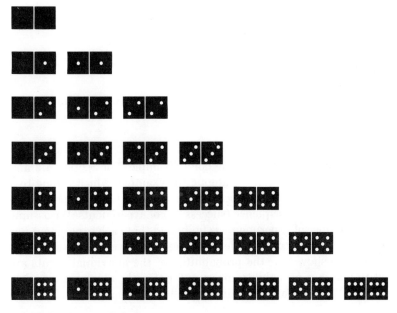

FIGURE 4.21   *Domino set*

    (d)   No double
    (e)   No double 4
    (f)   The sum of the dots is 7 exactly twice.

12.   In a 20-question multiple-choice examination, each question has
     four answers to choose from. All things being equal, what is the
     probability of correctly guessing

    (a)   Exactly 50 percent of the answers?
    (b)   Exactly 80 percent of the answers?

13.   A sequence of six traffic lights had originally been synchronized
     to enable a car driving at 45 mph to pass them without stopping.
     Due to increased traffic, the probability of passing a light with-
     out having to stop is $\frac{4}{5}$. Find the following probabilities:

    (a)   Passing all six lights without stopping
    (b)   Having to stop at each light
    (c)   Having to stop exactly twice

14.   Referring to Figure 4.21 (see also Exercise 11): Suppose five
     pairs of domino pieces are picked at random, one pair at a time,

each pair being replaced before the next one is picked. Find the probabilities of the following events:

(a) A pair of doubles each time
(b) A pair of doubles exactly twice
(c) A pair with one 6 on each piece exactly three times
(d) A pair for which the total of all dots is 3 exactly twice

4.6 FINITE STOCHASTIC PROCESSES

The repeated tossing of a single coin, Example 2 of Section 4.2 and Example 6 of Section 4.3, are among the examples which belong to a class of experiments called *stochastic*, or *random*. Loosely speaking, a *stochastic process* is a sequence (succession) of experiments. We assume that the number of experiments is finite and that each experiment has only a finite number of possible outcomes. What we want to do in this section is to show how the probabilities of the individual events are used to obtain a probability measure for the whole process (see Section 4.2). We illustrate this with the following example.

Example 1

The game in Figure 4.22 consists of branches $C_1, C_2, \ldots, C_{19}$ from 0 to one of the points $c_1, c_2, \ldots, c_{19}$. The probabilities of moving from one point to the next are given in the figure. A branch in this tree diagram is a stochastic process. It represents a sequence of moves or paths. We observe that

(1) The outcome of each trial (each move) is independent of the preceding trials.
(2) The probability of each move remains the same throughout the whole process.

Designating by A_j the path from 0 to a_j and by B_j the path from 0 to b_j, by C_j the path from 0 to c_j, we see that

$$P(C_1) = P(A_1) \cdot P(B_1|A_1) \cdot P(C_1|B_1) = \frac{1}{3} \times \frac{1}{3} \times \frac{1}{3} = \frac{1}{27}$$

$$P(C_{10}) = P(A_2) \cdot P(B_5|A_2) \cdot P(C_{10}|B_5) = \frac{1}{3} \times \frac{1}{2} \times \frac{1}{4} = \frac{1}{24}$$

and so on.

In this game the sample space S consists of the branches (paths)

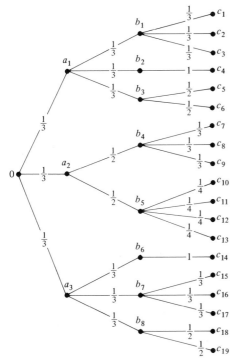

FIGURE 4.22

$C_1, \ldots, C_{19}$. To each of these branches (sample points) we assign the weight which is the product of the weights of its "sub-branches." The reader can readily verify that $P(S) = 1$ and that this gives a probability measure on the tree.

Example 2
Find the probability of reaching c_{13} or c_{19}.

SOLUTION

These are exclusive events (see Section 4.2), and hence

$$P(C_{13} \cup C_{19}) = P(C_{13}) + P(C_{19}) = \frac{1}{3} \times \frac{1}{2} \times \frac{1}{4} + \frac{1}{3} \times \frac{1}{3} \times \frac{1}{2} = \frac{7}{72}$$

Example 3
A ball-bearing manufacturer has four machines (M_1, M_2, M_3, M_4), each turning out 10^5 steel balls per week. A ball is acceptable if it

is within 0.0001 inch of exact measurement; from long experience, the following probabilities were assigned to each machine:

Machine	Probability for an unacceptable ball
M_1	$\dfrac{5}{1,000}$
M_2	$\dfrac{8}{1,000}$
M_3	$\dfrac{1}{1,000}$
M_4	$\dfrac{4}{1,000}$

The stochastic nature of this process is brought out clearly in Figure 4.23. The sample space S consists of eight branches (sample points), and a simple calculation shows that if the probability of each branch is the product of probabilities of "sub-branches," then $P(S) = 1$.

Example 4

What is the probability for an unacceptable ball from machine M_2?

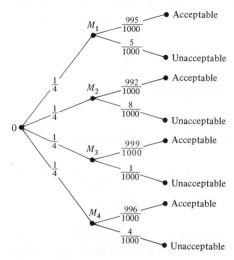

FIGURE 4.23

SOLUTION

The probability is

$$\frac{1}{4} \times \frac{8}{1,000} = \frac{1}{500}$$

Notice in these examples that the outcome at any stage (branch point) does not depend on previous outcomes. Those situations in which outcome is influenced by a preceding outcome are discussed in Chapter 7.

Example 5

The tree in Figure 4.24 gives data for a manufacturer of color television sets. What is the probability of getting a television set which is satisfactory after one service call?

SOLUTION

The cause for the service call could have been adjustment or repair. The probability for a satisfactory adjustment is

$$\frac{5}{20} \times \frac{7}{10} = \frac{35}{200}$$

The probability for a satisfactory repair is

$$\frac{2}{20} \times \frac{8}{10} = \frac{16}{200}$$

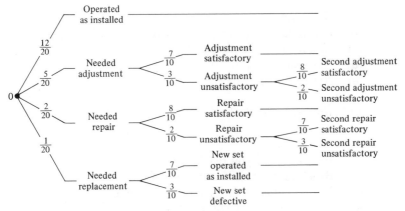

FIGURE 4.24 *Customer satisfaction with installed television sets*

Hence, the probability of satisfaction after one service call is

$$\frac{35}{200} + \frac{16}{200} = \frac{51}{200}$$

Example 6
Referring to Figure 4.24, what is the probability that a set will be unsatisfactory after a second repair?

SOLUTION

The probability is

$$\frac{2}{20} \times \frac{2}{10} \times \frac{3}{10} = \frac{3}{500}$$

Example 7
A pair of dice is rolled four times. What is the probability that the sums of the showing dots will be 2, then 4, then 6, then 8?

SOLUTION

Consult Exercise 7 in Section 4.1. The probabilities are, in turn, $\frac{1}{36}$, $\frac{3}{36}$, $\frac{5}{36}$, and $\frac{5}{36}$. Hence, the probability of the given outcome is

$$\frac{1}{36} \times \frac{3}{36} \times \frac{5}{36} \times \frac{5}{36} = \frac{75}{36^4}$$

EXERCISES

1. Referring to Example 1: Compute the probabilities of

    (a) Reaching c_5.
    (b) Reaching one of the points c_{10} or c_{11} or c_{12}.
    (c) Reaching one of the points c_7 or c_8 or c_9.
    (d) Not reaching c_1.
    (e) Not reaching c_{18} or c_{19}.

2. Referring to Example 1: Compute the conditional probabilities of

    (a) Reaching c_4 if you reached a_1.
    (b) Reaching c_7 or c_{10} if you reached a_2.
    (c) Not reaching c_{18} or c_{19} if you reached a_3.
    (d) Not reaching c_7 if you reached b_4.
    (e) Not reaching c_7 or c_8 or c_9 if you reached a_2.

3. Referring to Example 3: What is the probability of

    (a) An unacceptable ball from machine M_1 or machine M_2?
    (b) 1,800 unacceptable balls in a weekly production of all four machines?
    (c) 900 unacceptable balls in a weekly production of all four machines?

4. Referring to Example 5:

    (a) If a serviceman has to come twice, are you better off with a set which needed adjustment or a set that needed repair?
    (b) Are you better off with a set that needed service or with a set that needed replacement?

5. A human engineering firm used 120 persons selected at random to try out the operation of a newly designed home entertainment center in which a single panel contained controls for a record player, AM-FM radio, tape recorder, and a television set. Each participant was given four tests, each test consisting of 15 tasks. The results are listed in Table 4.6 in which "per" abbreviates *perfect performance* and "err" abbreviates *performance with errors.*

    (a) Draw a tree diagram for the sequence of four tests and indicate the probabilities between all branch points.
    (b) What is the probability for a perfect score on all four tests?
    (c) What is the probability of making an error on all four tests?

TABLE 4.6

per 80								err 40							
per 75				err 5				per 20				err 20			
per 75		err 0		per 4		err 1		per 5		err 15		per 10		err 10	
per 75	err 0	per 0	err 0	per 4	err 0	per 0	err 1	per 5	err 0	per 10	err 5	per 5	err 5	per 2	err 8

(d)   What is the probability of a perfect score on the fourth test
      if the participant made an error on the first test?
(e)   What is the probability of a perfect score on the fourth test
      if the participant made an error on a preceding test?

4.7 EXPECTED VALUE

When the possible outcomes of an experiment are numbers, it is often
useful to associate with the experiment a kind of average called its
expected value.

DEFINITION 1

Let the possible outcomes of an experiment be the numbers $a_1, a_2, \ldots,$
a_n with respective probabilities $p_1, p_2, \ldots, p_n$. The *expected value* of the
experiment is the number

$$E = a_1 p_1 + a_2 p_2 + \cdots + a_n p_n \qquad (20)$$

We shall illustrate this definition with a number of examples.

Example 1

A test in a class of 12 students resulted in the following grade
distribution:

Grade	60	70	80	90	100
Number of students	4	4	2	1	1

What is the expected value of the grade distribution?

SOLUTION

Let us put $a_1 = 60$, $a_2 = 70$, $a_3 = 80$, $a_4 = 90$, and $a_5 = 100$.
We note that the probabilities of these grades are, respectively,

$$p_1 = \frac{4}{12} \qquad p_2 = \frac{4}{12} \qquad p_3 = \frac{2}{12} \qquad p_4 = \frac{1}{12} \qquad p_5 = \frac{1}{12}$$

By Formula (20) the expected value is

$$E = 60 \times \frac{4}{12} + 70 \times \frac{4}{12} + 80 \times \frac{2}{12} + 90 \times \frac{1}{12} + 100 \times \frac{1}{12}$$
$$= 72.5$$

Example 2

A single die is rolled once. What is the expected value for the distribution of the sums of the showing dots?

SOLUTION

In this situation the possible outcomes are equiprobable, each having probability $\frac{1}{6}$. The expected value is therefore

$$E = 1 \times \frac{1}{6} + 2 \times \frac{1}{6} + 3 \times \frac{1}{6} + 4 \times \frac{1}{6} + 5 \times \frac{1}{6} + 6 \times \frac{1}{6}$$

$$= \frac{1}{6} \times (1 + 2 + 3 + 4 + 5 + 6) = 3.5$$

Observe that the expected value here is simply the arithmetic average of the six possible outcomes. In general, the expected value is always the arithmetic average when the possible outcomes are equiprobable.

Example 3

Suppose that a class of 40 students is graded on a pass-fail basis. For computation purposes, the number 1 is assigned to a pass grade, the number 0 to a fail grade. What is the expected value of the grade distribution if 28 students passed a given test?

SOLUTION

We note the following probabilities:

$$P(\text{pass}) = \frac{28}{40}$$

$$P(\text{fail}) = \frac{12}{40}$$

The expected value is therefore

$$E = 1 \times \frac{28}{40} + 0 \times \frac{12}{40} = \frac{28}{40}$$

Thus,

$$E = \text{proportion of students that passed}$$

The situation in this example is a dichotomy (division into two parts), and it is also worthwhile noting that

$$\text{sum of grades} = \text{number of students that passed}$$

It is apparent from the preceding examples that the expected value of an experiment should not be thought of as an outcome that will necessarily occur. In fact, the expected values in the examples were not even among the possible outcomes. The expected value has its meaning in experiments which are repeated a large number of times. Under these circumstances, the expected value represents an average for the experiment. Let us explain this with an example.

Example 4

Suppose that in a coin-toss game, you win 10 cents when Head comes up, and you lose 10 cents when Tail comes up. Since these two possible outcomes are equiprobable, the expected value for the game is

$$E = 10 \times \frac{1}{2} - 10 \times \frac{1}{2} = 0$$

Once more the expected value is not one of the possible outcomes of the game. Its interpretation is this: If the game is repeated a large number of times, then you should expect to break even, that is, your average winnings will be zero.

In general, a game with expected value zero is said to be *fair*; a game is said to be *favorable* if the expected value is positive, and *unfavorable* if the expected value is negative.

Example 5

In rolling a single die, you win 5 cents per dot when the number of dots is odd, and you lose 4 cents per dot when the number of dots is even. Is this game fair, favorable, or unfavorable?

SOLUTION

In this game your possible winnings are 5, 15, or 25 cents, and your possible losses are 8, 16, or 24 cents. Each possible outcome has probability $\frac{1}{6}$, and hence the expected value for the game is

$$E = (5 + 15 + 25) \times \frac{1}{6} - (8 + 16 + 24) \times \frac{1}{6} = -\frac{1}{2}$$

Since the expected value is negative, the game is unfavorable.

EXERCISES

1. Find the expected value for the grade distribution in the table below.

Student	A	B	C	D	E	F	G	H	I	J
Grade	70	70	60	80	70	80	60	70	70	90

2. In inspecting the adjustments on a batch of 100 television sets, a set is marked 0 if one or more adjustments are off, and it is marked 1 if all adjustments are correctly set. What is the expected value if

(a) All adjustments are correctly set?
(b) Each television has at least one adjustment which is off?
(c) Twenty-eight television sets have at least one adjustment which is off, but all other sets are perfectly adjusted?

3. In rolling a single die, you win 4 cents per dot if the number of dots is odd, and you lose 3 cents per dot if the number of dots is even. Is this game fair, favorable, or unfavorable?

4. Find the expected value for a coin toss if you win 2 cents when Head shows and you lose 3 cents when Tail shows.

5. Four coins are tossed. What is the expected value if

(a) The payoff is number of Heads minus number of Tails?
(b) The payoff is number of Tails minus number of Heads?
(c) You win 3 cents when exactly 3 Heads show, and lose 1 cent otherwise?
(d) You win 4 cents when exactly 3 Heads show, and lose 1 cent otherwise?

6. Four cards are drawn one at a time from a regular deck without replacement. You win eight dollars when all four cards are of the same color, and you lose one dollar otherwise. Is this game fair, favorable, or unfavorable?

5
LINEAR RELATIONS AND LINEAR PROGRAMMING WITH TWO VARIABLES

5

5.1 COORDINATES IN THE PLANE AND LINEAR EQUATIONS

In Section 0.1 we mentioned briefly the association of numbers with points on a straight line. This association will now be used to associate points in the plane with pairs of numbers. This is done by constructing a two-dimensional coordinate system: We select two perpendicular number lines intersecting in the origin 0 of each. One of these lines is customarily drawn horizontal and labeled the *x-axis*; the other line, which is vertical, is labeled the *y-axis*. The two axes are positioned as in Figure 5.1.

A system of two perpendicular number lines intersecting in the point 0 of each is called a *coordinate system*.

The terms *rectangular coordinate system* and *cartesian coordinate system* are also used. The plane containing an *xy*-coordinate system is called the *xy-plane* or *cartesian plane*.

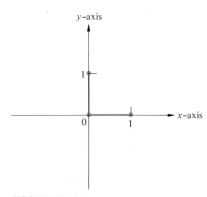

FIGURE 5.1

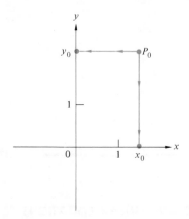

FIGURE 5.2

To explain the association of points with pairs of numbers, pick any point P_0 in the xy-plane and draw through P_0 perpendicular lines to the x- and y-axes (see Figure 5.2). These lines intersect the respective axes in points x_0 and y_0. Conversely, the perpendicular lines through x_0 and y_0 intersect in the unique point P_0. The point P_0 and the points x_0 and y_0 on the coordinate axes therefore completely determine each other, and we can thus put $P_0 = (x_0, y_0)$. We deduce that each point P in the plane determines a pair of numbers (x, y), and conversely. This implies that if (x, y) and (x', y') represent points in the plane, then

$$(x, y) = (x', y') \quad \text{when and only when} \quad x = x' \text{ and } y = y'$$

The pair (x, y) is called an *ordered pair*. The numbers x and y are called the *coordinates* of the point $P = (x, y)$, x being the *first* or *x-coordinate* and y the *second* or *y-coordinate*. The term *ordered pair* is justified in view of the fact that if

$$x \neq y \quad \text{then} \quad (x, y) \neq (y, x)$$

(See Figure 5.3.) The identification of points with ordered pairs of numbers can be summarized by saying:

A point in the cartesian plane is an ordered pair of real numbers.

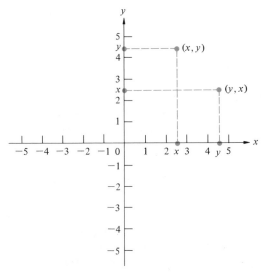

FIGURE 5.3

Example 1

A rectangle with sides parallel to the coordinate axes has two vertices: (1, 3) and (3, 1). Find the other two vertices.

SOLUTION

Consult Figure 5.4. One side of the rectangle lies one unit to the right of the y-axis and another side lies one unit above the x-axis.

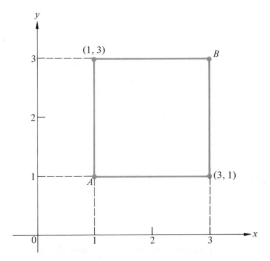

FIGURE 5.4

These sides intersect necessarily in the point $A = (1, 1)$. Likewise, the other two sides intersect in the point $B = (3, 3)$, and it turns out that the figure is actually a square.

Linear equations

The great power and utility of the concept of coordinates lies in the geometric interpretation that can be given to algebraic relations. Here we deal with the simplest of these, the so-called *linear equations*.

> **Example 2**
> Find the locus of points $P = (x, y)$ for which $x + y = 5$.
>
> *SOLUTION*
>
> The equation $x + y = 5$ can be solved for y to give $y = 5 - x$. We are thus seeking all points $P = (x, 5 - x)$. Plotting points for different values of x shows the locus to be a straight line through the points $(0, 5)$ and $(5, 0)$. (See Figure 5.5.)

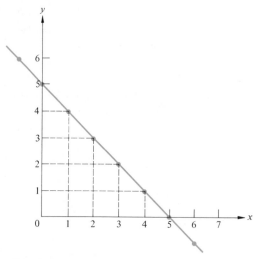

FIGURE 5.5

In general, the equation

$$ax + by = c$$

with not both a and b zero, is called a *linear equation* because the locus of points (x, y) satisfying it is a straight line, called *graph* of the equation. In Section 5.2 we shall show why the graph is a straight line. When graphing such an equation you should use the fact that

---

a straight line is uniquely determined by any two points on it.

Example 3

Find the graph of the equation $-2x + 5y = 3$.

SOLUTION

To find two convenient points, we take, in turn, $x = 0$ and $y = 0$. When $x = 0$, we get $5y = 3$, or $y = \frac{3}{5}$, and hence the point $(0, \frac{3}{5})$ lines on the line. When $y = 0$ we get $-2x = 3$, or $x = -\frac{3}{2}$, and we have for our second point $(-\frac{3}{2}, 0)$. Drawing the line containing these points gives the graph of $-2x + 5y = 3$. (See Figure 5.6.)

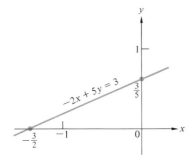

FIGURE 5.6

Example 4

Find the graph of the equation $2x = 3$.

SOLUTION

This equation can be written as

$$2x + 0 \cdot y = 3$$

It has a *unique* solution for x, namely, $x = \frac{3}{2}$, but it is satisfied by *all* values of y. The points satisfying the equation are therefore $(\frac{3}{2}, y)$, and two convenient points to plot are $(\frac{3}{2}, 0)$ and $(\frac{3}{2}, 1)$. We get a vertical line as drawn in Figure 5.7.

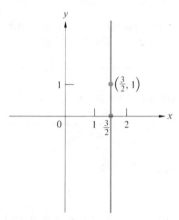

FIGURE 5.7

Example 5
Find the graph of the equation $5y = 1$.

SOLUTION

Writing this equation as

$$0 \cdot x + 5y = 1$$

we find that it is satisfied by all values x, but only by the unique value $y = \frac{1}{5}$. The points satisfying this equation are therefore $(x, \frac{1}{5})$, and two convenient points to plot are $(0, \frac{1}{5})$ and $(1, \frac{1}{5})$. The graph of the equation $5y = 1$ is a horizontal line (see Figure 5.8).

The results of Examples 2–5 are summarized in Table 5.1.

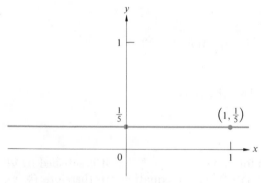

FIGURE 5.8

TABLE 5.1.   THE LINEAR EQUATION $ax + by = c$

Coefficients	Description of line
$a \neq 0, b \neq 0$	Line through $(0, c/b)$ and $(c/a, 0)$
$a \neq 0, b = 0$	Vertical line through $(c/a, 0)$
$a = 0, b \neq 0$	Horizontal line through $(0, c/b)$

EXERCISES

1. Plot the following points:

    (a)   (0, 3)   (3, 0)   (−3, 0)   (0, −3)
    (b)   (1, 4)   (−1, 4)   (−1, −4)   (1, −4)
    (c)   (−2, 0)   (−1, 1)   (0, 2)   (1, 3)
    (d)   (0, 5)   (2, 3)   (3, 2)   (5, 0)

2. What linear equation gives each of the following lines:

    (a)   The y-axis?
    (b)   The x-axis?
    (c)   The horizontal line through $(0, -2)$?
    (d)   The horizontal line through $(5, -2)$?
    (e)   The vertical line through $(1, 1)$?

3. Plot the graphs of the following linear equations:

    (a) $2x + 4y = 1$
    (b) $3x + 4y = -1$
    (c) $-x + 2y = -3$
    (d) $-x = 7$
    (e) $-y = 2$
    (f) $-\frac{1}{2}x + y = \frac{1}{3}$
    (g) $-0.5x - 0.1y = 1$
    (h) $\frac{1}{3}x - \frac{1}{4}y = 5$

4. Consider a right triangle with two sides parallel to the coordinate axes and of lengths three and four units, respectively. Find the vertices of all such triangles if one vertex is $(-2, 2)$ and the triangle crosses no coordinate axes.

5. Give the remaining two vertices of the square of sides of length 1 when two vertices are (1, 3) and (2, 4).

6. Consider the points $P = (-3, -4)$ and $Q = (x, y)$ and the line through these points. What can you say about the coordinates of Q if the line is

　(a) Parallel to the x-axis?
　(b) Parallel to the y-axis?
　(c) Parallel to the line through (1, 1) and (2, 0)?

7. Find the coordinates of a third point on the line through (0, 1) and (5, 6).

5.2 EQUATIONS OF LINES AND LINEAR FUNCTIONS

In Section 5.1 we considered the linear equation $ax + by = c$. When $b \neq 0$, we can solve for y to get $y = -(a/b)x + (c/b)$. Using the notation $m = -(a/b)$ and $d = (a/b)$, this equation becomes

$$y = mx + d \tag{1}$$

We shall now discuss this equation in some detail. The number m has an important geometric interpretation.

　Let (x_1, y_1) and (x_2, y_2) be any two distinct points satisfying Equation (1). (See Figure 5.9.) Then

$$y_1 = mx_1 + d \quad \text{and} \quad y_2 = mx_2 + d$$

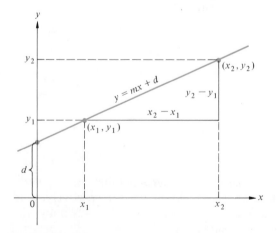

FIGURE 5.9

and consequently

$$\frac{y_2 - y_1}{x_2 - x_1} = \frac{(mx_2 + d) - (mx_1 + d)}{x_2 - x_1} = \frac{mx_2 - mx_1}{x_2 - x_1} = \frac{m(x_2 - x_1)}{x_2 - x_1}$$

$$= m$$

The number m represents the ratio of the change along the y-axis, divided by the corresponding change along the x-axis. This ratio is seen to be independent of the choice of points (x_1, y_1) and (x_2, y_2), and this confirms that (1) is the equation of a straight line. The number m is called the *slope* of the line and we have

$$\text{slope} = \frac{\text{change along vertical axis}}{\text{change along horizontal axis}}$$

Note:

The slope of a line is determined by any two distinct points (x_1, y_1) and (x_2, y_2) on it, provided $x_1 \neq x_2$. This restriction is necessary because the slope is $(y_2 - y_1)/(x_2 - x_1)$ and *we may never divide by zero.* Accordingly,

---

the slope of a vertical line is undefined.

If $x = 0$ in Equation (1), then $y = d$, and hence the point $(0, d)$ lies on the line $y = mx + d$. This point is called the *y-intercept.*

Slope-intercept formula

$y = mx + d$ is the equation of a line with slope m and y-intercept $(0, d)$.

Example 1

Find the equation of the line having slope $\frac{1}{2}$ and passing through the point $(0, -2)$.

SOLUTION

According to the given data, $m = \frac{1}{2}$ and $d = -2$. The equation of the line is therefore

$$y = \frac{1}{2}x - 2$$

(See Figure 5.10.)

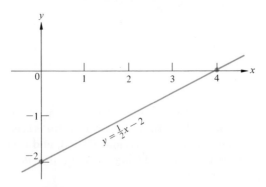

FIGURE 5.10

Now, if (x, y) and (x_0, y_0) are distinct points on a line with slope m, then

$$\frac{y - y_0}{x - x_0} = m$$

Multiplying both sides by $x - x_0$ gives

$$y - y_0 = m(x - x_0)$$

and this equation holds for *all* points (x, y) on the line.

Point-slope formula

$y - y_0 = m(x - x_0)$ is the equation of a line through the point (x_0, y_0) and having slope m.

Example 2
Find the equation of the line passing through the point $(1, 2)$ and having slope $-\frac{3}{2}$.

SOLUTION
Putting $(x_0, y_0) = (1, 2)$ and $m = -\frac{3}{2}$ gives the equation

$$y - 2 = -\frac{3}{2}(x - 1)$$

The line is given in Figure 5.11.

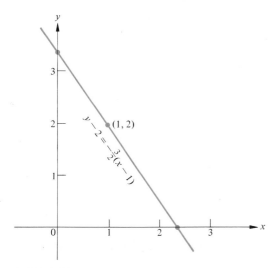

FIGURE 5.11

One more form for the equation of a line is often useful. It is this: If (x_0, y_0), (x_1, y_1), and (x, y) are three distinct points on a line with slope m, then

$$m = \frac{y - y_0}{x - x_0} = \frac{y_0 - y_1}{x_0 - x_1}$$

This formula is true for all points (x, y) for which $x \neq x_0$.

Two-point formula

$y - y_0 = \dfrac{y_0 - y_1}{x_0 - x_1}(x - x_0)$ is the equation of a line through the points (x_0, y_0) and (x_1, y_1).

Example 3

Find the equation of the line through the points $(-2, 2)$ and $(3, 4)$, and determine its slope.

SOLUTION

Putting $(x_0, y_0) = (-2, 2)$ and $(x_1, y_1) = (3, 4)$ gives for the equation of the line

$$y - 2 = \frac{2 - 4}{-2 - 3}[x - (-2)] \quad \text{or} \quad y - 2 = \frac{2}{5}(x + 2)$$

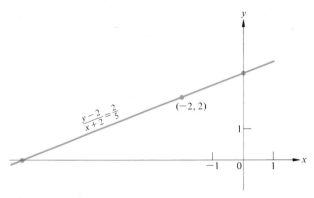

FIGURE 5.12

(See Figure 5.12.) The slope can be determined from the two given points, and hence

$$\text{slope} = m = \frac{2}{5}$$

Parallel lines

Consider two lines

$$y = mx + d_1 \quad \text{and} \quad y = mx + d_2$$

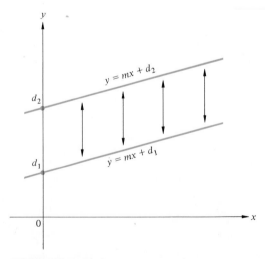

FIGURE 5.13

having the same slope m (see Figure 5.13). The relation

$$mx + d_2 = (mx + d_1) + (d_2 - d_1)$$

tells us that if the lines are not coincident $(d_2 - d_1 \neq 0)$, then all vertical line segments joining the two lines have the same length. One line is therefore obtained from the other through a vertical displacement. Hence,

---

lines having the same slope are *parallel* or *coincident*.

Linear functions

The relation $y = mx + d$ converts points from the x-axis into points on the y-axis. This is an example of a *function*. In general, a function is a relation which converts elements from one set into elements from another set in a unique manner. The special functions we are dealing with here are called *linear functions*.

DEFINITION 1

A *linear function* of one variable is a relation $y = mx + d$ that associates with each value x the unique value $mx + d$, where m and d are fixed numbers.

Terminology and notation

The letter x in the equation $y = mx + d$ represents an unspecified number; it is a *variable*. Also, y represents an unspecified number, and it, too, is a variable. However, y is computed from x: Each value of x determines the unique value $mx + d$, which is y. This is the reason for calling x an *independent variable*, y a *dependent variable*. The formula

$$y = L(x)$$

read "y equals L of x," tells us that y is a linear function of x. Formulas such as $y = M(x)$ and $y = N(x)$ are also used to represent linear functions.

Example 4

To explain the use of the functional notation $y = L(x)$, consider the function $y = 2x + 1$. If we put

$$L(x) = 2x + 1$$

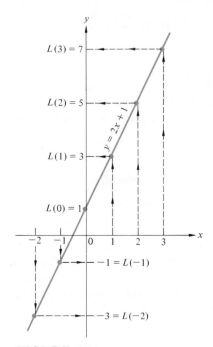

FIGURE 5.14

then

$$L(-2) = 2 \cdot (-2) + 1 = -3$$
$$L(-1) = 2 \cdot (-1) + 1 = -1$$
$$L(0) = 2 \cdot 0 + 1 = 1$$
$$L(3) = 2 \cdot 3 + 1 = 7$$

and so on (see Figure 5.14).

Example 5

A consultant has a fixed overhead (rent, utilities, etc.) of $450 per month. His fee is $120 per day. What will be his profit if he works *x* days per month?

SOLUTION

The consultant's income from *x* days of work is

$$I(x) = 120x$$

His monthly expenses are

$$E(x) = 450$$

His profit is therefore

$$I(x) - E(x) = 120x - 450$$

The graphs of $y = I(x)$ and $y = E(x)$ are plotted in Figure 5.15. The value x_0 for which $I(x_0) = E(x_0)$ is called the *break-even point*. This is the number of days the consultant has to work just to recover his overhead. His profit is therefore negative (loss) for $x < x_0$ and positive (gain) for $x > x_0$. The break-even point in this example is $x_0 = \frac{450}{120} = 3.75$.

Break-even analysis is frequently used in business. We shall discuss it again in Section 5.3.

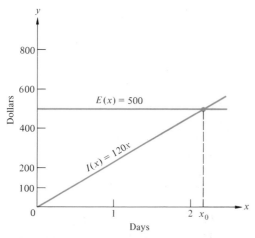

FIGURE 5.15

EXERCISES

1.   For each pair of points listed, find the slope of the line determined by it. When the line is vertical, write "m is undefined."

(a) $(-1, 1); (1, -1)$
(b) $(0, 0); (1, 2)$
(c) $(0, -1); (-1, 0)$
(d) $(a, b); (2a, b)$

(e) (1, 4); (1, 6)
(f) (1, 1); (5, 1)

In Exercises 2–8 find equations for the given lines.

2. Passing through (2, 3) and having slope 4
3. Having slope -1 and y-intercept (0, -2)
4. Passing through the points $(-2, -2)$ and (3, 4)
5. Passing through the point (1, 5) and parallel to the line

$$y - 3 = 2(x - 1)$$

6. Passing through the point (0, -3) and parallel to $-2x + 3y = 7$
7. Having y-intercept (0, -3) and passing through (4, 0)
8. Passing through the points (1, -1) and (1, 7)
9. We know that any *two* points determine a line. In each case below, decide if the points are collinear.

(a) $(-1, 1)$, (1, 2), (11, 10)
(b) $(-4, -3)$, $(2, -1)$, $(-10, -7)$
(c) (a, b), $(a + 1, b + 1)$, $(a - 1, b - 1)$

10. A real estate broker has a fixed overhead of $800 per month. His commission is 6 percent of the value of real estate he sells. Following the procedure in Example 5, plot his income and overhead graphs, and indicate the break-even point.

11. Find the indicated values of the given linear functions.

(a) $L(x) = 3x - 1$ $L(0)$ $L(1)$ $L(-1)$
(b) $M(x) = -2x + 5$ $M(\frac{5}{2})$ $M(1)$ $M(-1)$
(c) $N(x) = 3$ $N(0)$ $N(-2)$ $N(3)$
(d) $Q(x) = 4(x - 1)$ $Q(1)$ $Q(-1)$ $Q(-2)$
(e) $R(x) = -4x - 4$ $R(1)$ $R(-1)$ $R(\frac{1}{4})$
(f) $S(x) = x$ $S(-5)$ $S(5)$ $S(7)$

5.3 SYSTEMS OF TWO LINEAR EQUATIONS

Two straight lines in a plane, when they are not parallel or coincident, intersect in a single point. In this section we shall develop a method for finding the point of intersection, and then we shall apply this to break-even analysis.

Example 1

Find the point of intersection of the two lines

$$2x + 3y = 4$$
$$3x + 8y = -1$$
(2)

SOLUTION

The two equations constitute a *system* of *linear equations* in the unknowns x and y. The problem is that of finding a point (x_0, y_0) in the xy-plane such that

$$2x_0 + 3y_0 = 4$$
$$3x_0 + 8y_0 = -1$$

(See Figure 5.16.)

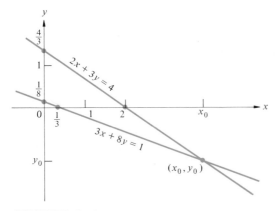

FIGURE 5.16

In the procedure below the idea is to eliminate *one* of the variables, solve for the other variable, and then use this to solve for the eliminated variable. The procedure is illustrated schematically with a flow diagram (see Figure 5.17).

We find that the solution of the system of equations is $(5, -2)$, and this is the point of intersection of the corresponding lines.

Remark

The variable you choose to eliminate first is unimportant. With a little experience, however, you will pick the order which requires the least number of computations.

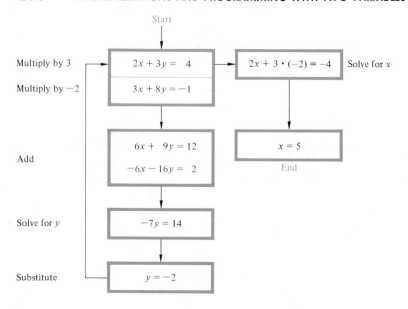

FIGURE 5.17   *Schematic diagram for solving the system of linear equations (2)*

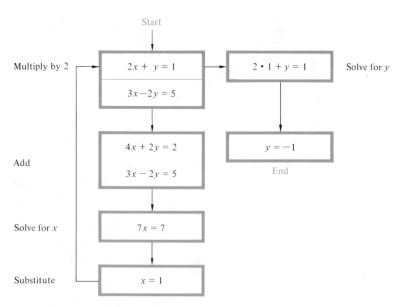

FIGURE 5.18   *Schematic diagram for solving the system of linear equations (3)*

Example 2

Find the solution of the system of linear equations

$$2x + y = 1$$
$$3x - 2y = 5$$
(3)

SOLUTION

You should be able to follow the flowchart in Figure 5.18. The solution of the system is found to be $(1, -1)$ (see Figure 5.19).

From the geometric interpretation of the solution of a system of linear equations as the point of intersection of two lines, we know that

(1) If a solution exists, it is unique.
(2) No solution may exist.

When no solution exists, this will show itself as we try to find one.

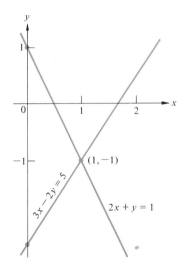

FIGURE 5.19

Example 3

Find the solution of the system

$$x - \frac{1}{2}y = 1 \quad \text{and} \quad -6x + 3y = -12$$

SOLUTION

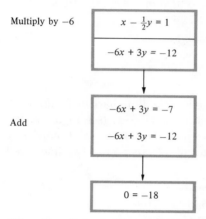

Multiply by -6

$$x - \tfrac{1}{2}y = 1$$

$$-6x + 3y = -12$$

Add

$$-6x + 3y = -7$$

$$-6x + 3y = -12$$

$$0 = -18$$

The absurdity of this conclusion tells us that the equations are *inconsistent* and the system has no solution. Indeed, we find that the equations give parallel lines (see Figure 5.20).

We now apply our newly acquired knowledge to two types of problems in business and economics.

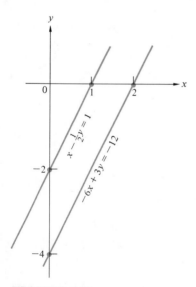

FIGURE 5.20

Break-even analysis

This concept was introduced in Example 5 of Section 5.2, and it is summarized graphically in Figure 5.21, where a cost function is compared with a revenue function. The cost usually consists of a fixed overhead, which is independent of any production or service, and a variable cost which depends on the amount of goods produced or services rendered. The revenue is the total cash income, and the break-even point occurs when revenue equals cost.

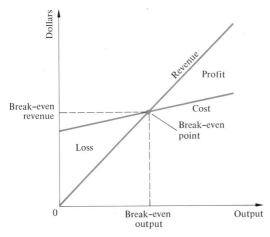

FIGURE 5.21

Example 4

A manufacturer of magnetic recording tape sells his product at an average of $2.4 per 1,000 feet. Production costs consist of

A fixed overhead of $660 per month
Variable production cost of $0.8 per 1,000 feet

Find

(1) How many units of 1,000 feet of tape must be sold per month to break even?
(2) What is the revenue at the break-even point?

SOLUTION

If the manufacturer sells x units of 1,000 feet, his revenue (total income from the sale of the product) and production cost are

$$R(x) = 2.4x \text{ dollars} \qquad \text{(revenue)}$$
$$C(x) = 0.8x + 660 \text{ dollars} \qquad \text{(cost)}$$

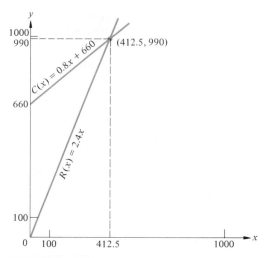

FIGURE 5.22

(See Figure 5.22.) Putting $y = R(x)$ and $y = C(x)$, we write these equations as

$$\begin{align} y - 2.4x &= 0 \\ y - 0.8x &= 660 \end{align} \tag{4}$$

A solution of this system of equations will yield the answers to both questions. We proceed as in the earlier examples.

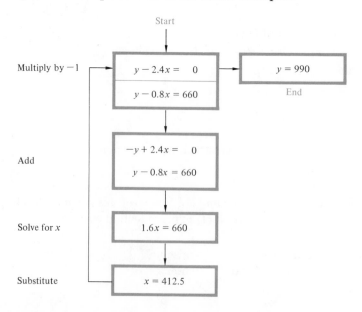

Thus, the lines in (4) intersect in the point (412.5, 990). The break-even point is 412.5, and to achieve it, revenue must be $990.

Market equilibrium

As a second application we consider the so-called market equilibrium for a product in economic theory.

Consider a market situation in which demand for a given commodity depends on its price: the higher the price, the lower the demand; the lower the price, the higher the demand. From the manufacturer's point of view, the commodity will not be produced if the price is zero, but the manufacturer can be expected to react to price and to increase production with rising price. This relation, the *price-supply* relation, clearly interacts with the *price-demand* relation. In many situations the price-demand and price-supply relations can be taken to be *linear*, and then the relations can be presented graphically as in Figure 5.23. As is customary in economics, price is given in terms of supply and in terms of demand (i.e., price is the "dependent variable").

The situation as set up here is an example of a *model*. An economist could analyze a given market situation, using such linear relations, and see if this gives usable answers. The point of intersection of the lines in Figure 5.23 is the point where supply equals demand, and this is the *market equilibrium* for the commodity at the *equilibrium price* p_0. The point x_0 is called the *equilibrium quantity*.

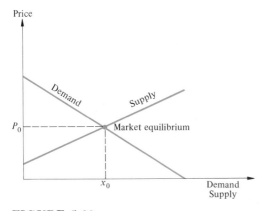

FIGURE 5.23

Example 5

A weekly market demand for a commodity is $q_d = 12 - 5p$ and the weekly market supply is $q_s = 2 + 3p$, where p is the price. Find the equilibrium price and quantity.

SOLUTION

For simplicity we write the two equations as

$$x + 5p = 12$$
$$x - 3p = 2$$

and we proceed to find the solution of this system.

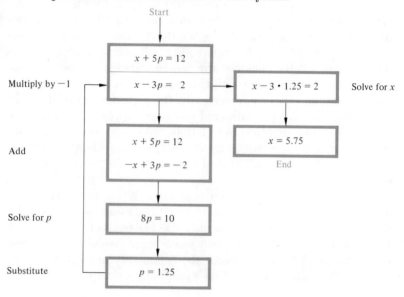

The equilibrium price is seen to be $p = 1.25$ and the equilibrium quantity is 5.75. The situation is given in Figure 5.24.

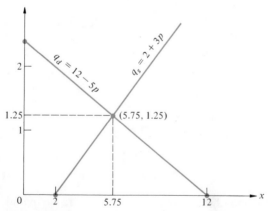

FIGURE 5.24

EXERCISES

Solve the systems of linear equations in Exercises 1–12 below. When no solution exists, state so.

1. $x + y = 1$
   $x - y = 1$

2. $x + 1 = 0$
   $y + 1 = 0$

3. $x + 3y = 6$
   $-x + 2y = -1$

4. $2x - y = -2$
   $-3x + 2y = 5$

5. $x - y = 0$
   $x + y = 0$

6. $\frac{1}{2}x - y = 1$
   $2x - y = 1$

7. $-\frac{1}{2}x + y = 7$
   $x + \frac{1}{3}y = -1$

8. $2x - 3y = 5$
   $x \quad\quad = -2$

9. $7x + 5y = 3$
   $-x + 2y = 0$

10. $x + 2y = 4$
    $2x - y = 3$

11. $3x - 4y = 6$
    $x + y = 9$

12. $3x + 2y = 5$
    $9x + 6y = -1$

13. A pen manufacturer has a fixed overhead of $180 per week and a variable cost of $0.5 per pen manufactured. If the pens are sold by the manufacturer for $0.9 each,

   (a) How many pens must be sold per week to break even?
   (b) What is the revenue at the break-even point?
   (c) What is the manufacturer's profit if he sells 5,000 pens in four weeks (profit = revenue − cost)?

14. A bottle manufacturer can produce 12-ounce bottles at the cost of 5 cents each; his fixed monthly overhead is $720. If he installs new equipment, his fixed monthly overhead will increase to $1,200, but the cost per 12-ounce bottle will reduce to 2 cents. If he sells his bottles for 8 cents each,

   (a) How many bottles must be sold per month to break even with the old equipment?
   (b) How many bottles must be sold per month to break even with the new equipment?
   (c) What is the manufacturer's profit in half a year with the

old equipment and new equipment if he sells a total of 250,000 bottles?

15. A person going into business has a choice between two manufacturing processes for a given product. The cost functions are

$C_1(q) = 0.4q + 125$ dollars
$C_2(q) = 0.1q + 500$ dollars

where q stands for the number of units produced, and $125 and $500 are fixed monthly costs.

(a) Which process should he choose if the expected average sale is 4,000 units per month at $0.7 each?
(b) Which process should he choose if the expected average sale is 1,000 units per month at $0.7 each?

16. Referring to Exercise 15: What is the minimum average monthly sale at $0.7 per unit which will justify the second manufacturing process?

In Exercises 17–20, find the equilibrium price and quantity.

17. $q_d = 1200 - 10p$
    $q_s = 500 + p$

18. $q_d = 500 - 0.1p$
    $q_s = 120 + 0.01p$

19. $q_d = 10 - 0.9p$
    $q_s = 3 + 0.5p$

20. $q_d = 7 - \frac{1}{2}p$
    $q_s = -1 + 3p$

5.4 SYSTEMS OF LINEAR INEQUALITIES

Linear inequalities, like linear equations, have a geometric interpretation in terms of a *graph*. Here, however, the graph is a region instead of a line. We shall explain how to graph linear inequalities by means of the following examples.

Example 1
Sketch the graph of the inequality $y - x \geq 0$.

SOLUTION

The graph of this inequality, which can be written as $y \geq x$, consists of all points (x, y) in the plane for which $y = x$ or $y > x$. Drawing the line $y = x$, pick a point (x_1, y_1) lying above it (see Figure 5.25). It is observed that $y_1 > x_1$. On the other hand, if (x_2, y_2) lies be-

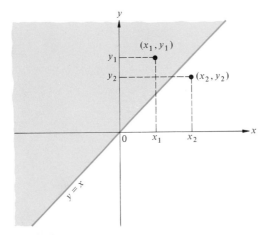

FIGURE 5.25

low the line $y = x$, then the figure shows that $y_2 < x_2$. The graph is therefore the shaded area including the line $y = x$.

Example 2
Sketch the graph of the inequality $y - \frac{1}{2}x < 1$.

SOLUTION
Writing this inequality in the equivalent form

$$y < \frac{1}{2}x + 1$$

we observe that any point (x, y) in the plane must satisfy exactly one of the following relations:

$$y < \frac{1}{2}x + 1$$

$$y = \frac{1}{2}x + 1$$

$$y > \frac{1}{2}x + 1$$

Drawing the line $y = \frac{1}{2}x + 1$, we see that, as in Example 1, $y < \frac{1}{2}x + 1$ for all points (x, y) lying below this line, whereas $y > \frac{1}{2}x + 1$ for all points lying above it. The graph of the inequality $y < \frac{1}{2}x + 1$ consists therefore of the portion of the plane lying below the line $y = \frac{1}{2}x + 1$ and not on it (see Figure 5.26).

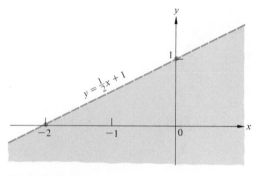

FIGURE 5.26

Example 3

In what region of the plane does the following system of inequalities hold?

$$y - x \geq 0$$

$$y - \frac{1}{2}x < 1$$

SOLUTION

The first inequality holds in the shaded region in Figure 5.25, and the second inequality holds in the shaded region in Figure 5.26. Superimposing these two figures gives the area in which both inequalities hold (see Figure 5.27). The shaded area is called the *graph* of the system of inequalities.

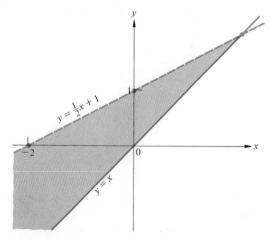

FIGURE 5.27

Example 4

Sketch the graph of the system of inequalities

$$y + 3x \geq 3$$
$$y + x \geq 2$$
$$8y + x \geq 4$$

SOLUTION

Reasoning as in Example 1 shows that the graph of the in-equality $y + 3x \geq 3$ consists of the points (x, y) lying on or above the line $y + 3x = 3$ [see Figure 5.28(a)]. Likewise, for the graph of $y + x \geq 2$, we take the region determined by the line $y + x = 2$ and lying above it. Superimposing this on Figure 5.28(a) gives the region in Figure 5.28(b). Finally, superimposing the region above the line $8y + x = 4$ gives the shaded area in Figure 5.28(c). This is the graph of the system of inequalities.

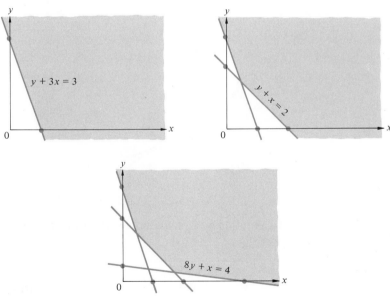

FIGURE 5.28

Example 5

Sketch the graph of the system of inequalities

$$4y + x \geq 4$$
$$y - x \leq 1$$
$$4y + 3x \leq 18$$

SOLUTION

Draw the lines

$$
\begin{aligned}
4y + \quad x &= \ 4 \\
y - \quad x &= \ 1 \\
4y + 3x &= 18
\end{aligned}
$$

(See Figure 5.29.) We know from Examples 1–4 that each point (x, y) satisfying the above system of inequalities must lie *on* or *above* the line $4y + x = 4$, *on* or *below* the line $y - x = 1$, and *on* or *below* the line $4y + 3x = 18$. Hence, the graph is the shaded area bounded by these lines.

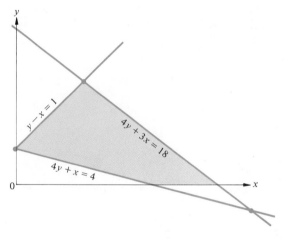

FIGURE 5.29

We conclude this section with some applications of systems of linear inequalities.

Example 6

A cheese product is required to have the following composition:

Butterfat: at least 27 percent
Dairy Products: at least 30 percent
Moisture: at least 10 percent

What combinations of butterfat and dairy products will satisfy these requirements?

SOLUTION

Let the percentage of butterfat in the product be x, the percentage of dairy products be y (the dairy products are assumed to contain no butterfat). Then we must have the inequalities

$$x \geq 27$$
$$y \geq 30 \tag{5}$$
$$x + y \leq 90$$

The graph of this system of inequalities is given in Figure 5.30. The shaded area is the *set of solutions* of the system; that is, each pair of values (x, y) satisfies these inequalities if and only if it lies in the shaded area (or on the lines bounding it). Hence, we have found the relative percentages we wanted.

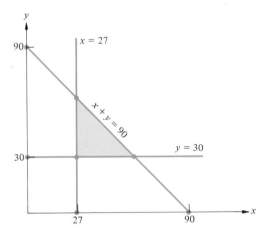

FIGURE 5.30

Example 7

An animal feed is composed of meat by-products, enriched cereals, and a filler. The composition of these compounds is given in Table 5.2. One pound of the feed is required to contain the following minimum amounts:

Protein:	150 grams
Fat:	45 grams
Carbohydrates:	400 grams

What mixtures of meat by-products and cereal will satisfy these constraints?

TABLE 5.2. WEIGHTS ARE GIVEN IN GRAMS PER OUNCE

	Meat by-products	Cereals
Protein	20	50
Fat	15	5
Carbohydrates	40	320

SOLUTION

Let the amount of meat by-products in 1 pound of feed be x, the amount of cereal be y. Then

$$x + y \leq 16 \qquad (1 \text{ lb.} = 16 \text{ oz.})$$
$$x \geq 0$$
$$y \geq 0$$

According to the given data, the following additional inequalities must hold:

$$20x + 50y \geq 150$$
$$15x + 5y \geq 45$$
$$40x + 320y \geq 400$$

The mixture will meet the constraints for amounts x and y for which (x, y) lies in the solution-set of the preceding two systems of inequalities (see Figure 5.31).

Example 8

A firm is producing two models of tape recorder heads, T_1 and T_2, to which it can allocate the following facilities on a weekly basis:

Casting: 40 hours
Machining: 180 hours
Finishing: 80 hours

The average amount of time that has to be spent on each model is given in Table 5.3. Under these constraints, what are the quantities of T_1 and T_2 that the firm can produce per week?

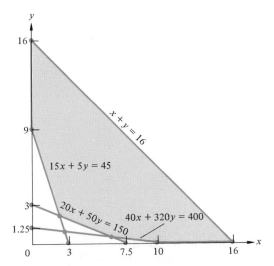

FIGURE 5.31

SOLUTION

Suppose the firm decides to produce x units of model T_1 and y units of model T_2. Then the data in Table 5.3 give the following inequalities:

$$x \geq 0$$

$$y \geq 0$$

$$\frac{1}{4}x + \frac{1}{6}y \leq 40$$

$$\frac{1}{2}x + \ y \leq 180$$

$$\frac{1}{6}x + \frac{1}{2}y \leq 80$$

TABLE 5.3

	Hours per unit	
	Model T_1	Model T_2
Casting	1/4	1/6
Machining	1/2	1
Finishing	1/6	1/2

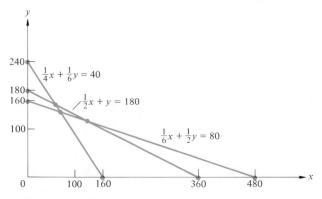

FIGURE 5.32

The quantities x and y are possible under the above constraints only if (x, y) lies in the solution-set of these inequalities (see Figure 5.32). You will observe in the figure that machining capacity is never reached, and hence it is not really a constraint.

EXERCISES

Sketch the graphs of the inequalities in Exercises 1–12.

1. $x > 1$

2. $y > -1$

3. $2y - 4x > 3$

4. $2x + 3y \leq 1$

5. $y - x > -\dfrac{1}{2}$

6. $-5y + 4x \leq 12$

7. $-1 \leq y - \dfrac{1}{2}x$

8. $y \leq 2$

9. $\dfrac{1}{2}y + x \leq 2$

10. $x + y \leq 1$

11. $2x - 3y \leq 0$

12. $x - y \leq 4$

Sketch the graphs of the system of inequalities in Exercises 13–22.

13. $x > 1$
    $y > -1$

14. $x + y < 0$
    $x - y < 0$

15. $2x + 3y \geq 1$
    $-2x + 3y \geq 1$

16. $y - x \geq 0$
    $y + x \leq 3$
    $x \geq 0$

17. $6y - x \leq 19$
    $-5x + 4y \leq 12$
    $y \geq 0$

18. $x \geq 0$
    $x \leq 3$
    $y \geq 0$
    $y \leq 5$

19. $0 \leq x - y$
    $0 \leq x + y$

20. $x \geq 0$
    $y \geq 0$
    $x + y \leq 16$
    $3x + 6y \geq 18$
    $4x + 5y \geq 20$

21. $x \geq 0$
    $y \geq 0$
    $x + y \leq 4$
    $x + 2y \leq 5$

22. $x \geq 0$
    $y \geq 0$
    $3x + 4y \leq 16$
    $2x + y \leq 8$
    $x - y \leq 2$

23. A pet food is required to have the following composition:

    Meat and meat by-products: at least 25 percent
    Cereals: at least 50 percent
    Moisture: at least 15 percent

    What relative percentages of meat and its by-products and cereals will meet these requirements?

24. A fertilizer is a mixture of organic and synthetic compounds with the composition listed in Table 5.4. What mixture will guarantee 9 percent nitrogen, 5 percent phosphorus, and 15 percent potassium?

TABLE 5.4

	Organic compound, %	Synthetic compound, %
Nitrogen	6	12
Phosphorus	6	4
Potassium	10	20

5.5 LINEAR PROGRAMMING

Many important decision-making problems in business involve the finding of the largest (maximum) or smallest (minimum) value of a linear function, when the variables are nonnegative and restricted to the solution-set of a system of inequalities. Thus, the problem is one of maximizing or minimizing a linear function over a certain set of points, and the technique is called *linear programming*. In this section we shall deal with problems involving two variables, and we begin with an example.

Example 1

A firm manufactures two models, A and B, of a small electric appliance. The production costs and the weekly constraints are given in Table 5.5. How can the firm maximize its profit from these two items if the wholesale price it charges per unit is $22.5 for model A and $30.5 for model B?

TABLE 5.5

	Cost per unit, dollars		Maximum weekly allocation, dollars
	Model A	Model B	
Parts	1.5	2.5	3,000
Wiring	12	8	15,000
Assembly	3	8	8,700

SOLUTION

Let us begin by determining the number of units that can be manufactured for each model under the above constraints. If x is the number of units of model A, and y is the number of units of model B, then, by the data in Table 5.5,

$$x \geq 0$$
$$y \geq 0$$
$$1.5x + 2.5y \leq 3{,}000$$
$$12x + 8y \leq 15{,}000$$
$$3x + 8y \leq 8{,}700$$

The quantities x and y satisfying this system of inequalities are found from the solution-set (see Figure 5.33).

Now, the firm's cost per unit is $16.5 for model A, and $18.5 for model B. Hence, the profit per unit is $22.5 - 16.5 = 6$ dollars for model A, and $30.5 - 18.5 = 12$ dollars for model B. The total profit

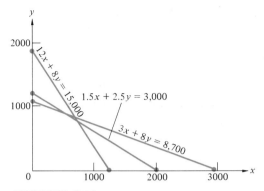

FIGURE 5.33

from the sale of x units of model A and y units of model B is therefore $6x + 12y$, and this is the quantity we want to maximize. For different values of f, the lines

$$f = 6x + 12y$$

are parallel, since they have the same slope. In Figure 5.34, which is an enlargement of the solution-set in Figure 5.33, the lines for several values of f have been drawn. The direction of increasing f, which is the direction of maximizing $6x + 12y$, is shown by arrows. We see intuitively that a maximum will be reached by the farthest line in the di-

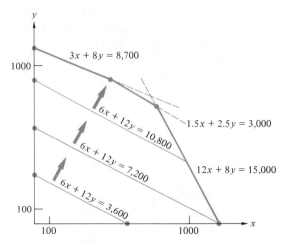

FIGURE 5.34

rection of the arrows which still intersects the shaded region. This is the line passing through the point of intersection of the lines:

$$1.5x + 2.5y = 3,000$$
$$3x + 8y = 8,700$$

This point is found with the technique discussed in Section 5.3 for solving a system of two linear equations.

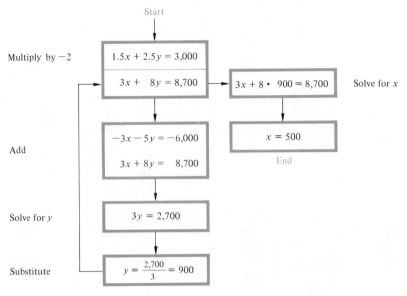

The point of intersection is therefore (500, 900). Analyzing these data, we find that the firm reaches maximum profit when it produces 500 units of model A and 900 units of model B. The firm's revenue and profit from these two items are computed in Table 5.6.

We have seen in this example that a maximum value was reached at a corner, or vertex, of the set of solutions of the system of inequali-

TABLE 5.6

	Model A (500 units)	Model B (900 units)	Total
Cost	8,250	16,650	24,900
Revenue	11,250	27,450	38,700
Profit	3,000	10,800	13,800

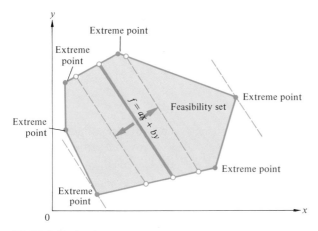

FIGURE 5.35

ties. We explain below why this is true in general, but first we introduce some terminology.

In the terminology of linear programming, the function to be maximized or minimized is called the *objective function*. Since a solution is feasible only if it lies in the solution-set of the constraints(inequalities), this set is called the *feasibility set* or *feasibility region*. The shaded region in Figure 5.34 has the property that a line segment joining any two points on its boundary lies entirely in it or on its boundary. A set with this property is said to be a *polygonal convex set*. The vertices (points of intersection of two sides) are called the *extreme points* of the set (see Figure 5.35).

We now make the following observation:

Let S be a polygonal convex set in the plane having a finite number of extreme points (vertices). Let $f = ax + by$ be any line intersecting S, and consider the line segment lying in S (see Figure 5.35). From the behavior of lines we know that

    (1)   Different values of f give parallel lines.
    (2)   Moving from $f = ax + by$ in one direction increases f, whereas moving in the opposite direction decreases f.

Hence, we see intuitively that $ax + by$ will be maximized and minimized at an extreme point of S. The procedure for maximizing or minimizing an objective function is therefore this:

    (1)   Find the extreme points of the feasibility set:

$$(x_1, y_1), \quad (x_2, y_2), \ldots, \quad (x_n, y_n)$$

(2) Substitute the values (x_i, y_i) found in (1) into $ax + by$. The set of numbers $\{ax_1 + by_1, ax_2 + by_2, \ldots, ax_n + by_n\}$ has a largest and a smallest element. These are, respectively, the maximum and minimum of the objective function.

Remark 1

There may be more than one extreme point which maximizes or minimizes an objective function. In fact, if the objective function is parallel to a side of the feasibility set, any point on this side may be a solution.

Remark 2

A practical difficulty arises when the feasibility set has a large number of extreme points. A simple diagram is often helpful, since it may indicate the likely candidates for a solution. In some cases, in fact, a graphical solution is simpler to obtain. When the number of constraints is very large, it may even be difficult to find all the extreme points.

Example 2

In Example 1, suppose that, due to demand, the firm is able to get \$30.5 for each unit of each model. What production under the given constraints will maximize profit?

SOLUTION

The firm's profit is $30.5 - 16.5 = 14$ dollars on each unit of model A and $30.5 - 18.5 = 12$ dollars on each unit of model B. Its profit from the sale of x model A units and y model B units is therefore $14x + 12y$. In Figure 5.36 we plotted the lines $f = 14x + 12y$

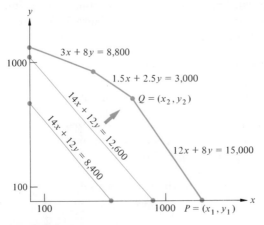

FIGURE 5.36

for two values of f, and we noted the direction of increase (the direction of increase is, of course, known from Example 1). The extreme points $P = (x_1, y_1)$ and $Q = (x_2, y_2)$ identified in Figure 5.36 are seen to be candidates. P is particularly easy to find; setting $y = 0$ in $12x + 8y = 15{,}000$ gives $x = 1{,}250$, and hence

$$P = (1{,}250,\ 0)$$

The point Q is found as usual.

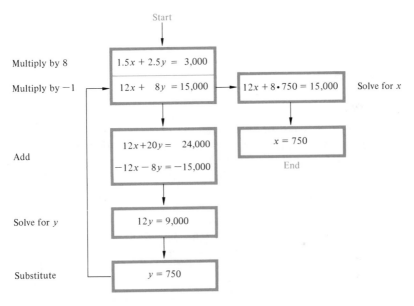

Hence, $Q = (750,\ 750)$, and a substitution in $14x + 12y$ gives the following:

$$P: 14 \times 1{,}250 + 12 \times \quad 0 = 14 \times 1{,}250 = 17{,}500$$
$$Q: 14 \times \quad 750 + 12 \times 750 = 16 \times \quad 750 = 19{,}500$$

We see that profit is maximized when 750 units of each model are produced.

Example 3
Consider the feasibility set in Figure 5.37. Minimize the objective function

$$f = \frac{1}{6}x + y$$

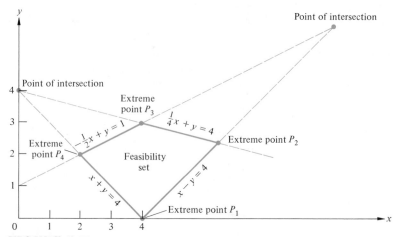

FIGURE 5.37

SOLUTION

Here it is easy to substitute the extreme points $P_j = (x_j, y_j)$ into $\frac{1}{6}x + y$. We see that $\frac{1}{6}x + y$ is minimized at both P_1 and P_2, and hence it is minimized for each point (x, y) on the line segment joining P_1 and P_2. We observe in passing that the objective function $\frac{1}{6}x + y$ is maximized at P_4.

P_j	(2, 2)	(8, 1)	(10, 5)	(7, 9)	(3, 8)	(1, 6)
$\frac{1}{6}x_j + y_j$	$\frac{14}{6}$	$\frac{14}{6}$	$\frac{40}{6}$	$\frac{61}{6}$	$\frac{51}{6}$	$\frac{37}{6}$

Example 4

Consider the set of solutions of the inequalities

$$x + y \geq 4$$

$$-\frac{1}{2}x + y \leq 1$$

$$\frac{1}{4}x + y \leq 4$$

$$x - y \leq 4$$

Subject to these constraints, maximize the objective function $f = 50 - 2x + y$.

SOLUTION

We begin with a caution: The feasibility set is determined by four lines, any two of which intersect. There is a total of $\binom{4}{2} = 6$ intersections, but only four of these are extreme points of the feasibility set. We must, therefore, begin by finding the intersections which give extreme points. This is easily done with a simple diagram.

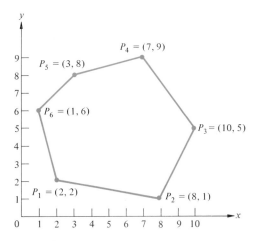

FIGURE 5.38

From Figure 5.38 we find the following extreme points:

$P_1 = (4, 0)$

$P_2 = \left(\dfrac{32}{5}, \dfrac{12}{5}\right)$

$P_3 = (4, 3)$

$P_4 = (2, 2)$

The substitution into $50 - 2x + y$ is readily carried out, and we see that the objective function $50 - 2x + y$ is maximized at $P_4 = (2, 2)$.

P_j	$(4, 0)$	$\left(\dfrac{32}{5}, \dfrac{12}{5}\right)$	$(4, 3)$	$(2, 2)$
$50 - 2x_j + y_j$	42	39.6	45	48

The method we introduced here for solving linear programming problems in two variables is simple and effective, at least when the number of extreme points is not too large. More sophisticated problems and more variables require more sophisticated methods.

EXERCISES

1. Referring to Figure 5.34: Maximize the following objective functions:

   (a) $f = 2x + y$
   (b) $f = x + y$
   (c) $f = y$
   (d) $f = x + 10y$

2. Referring to the feasibility set in Figure 5.37:

   (a) Minimize $f = x$
   (b) Minimize $f = x + y$
   (c) Minimize $f = 6x - y$
   (d) Maximize $f = x$
   (e) Maximize $f = x + y$
   (f) Maximize $f = 6x - y$

3. Referring to the feasibility set in Figure 5.32:

   (a) Minimize $f = y$
   (b) Minimize $f = y - x$
   (c) Maximize $f = x + y$
   (d) Maximize $f = \dfrac{1}{20}x + y$

4. Referring to the feasibility set in Figure 5.31:

   (a) Minimize $f = y$
   (b) Minimize $f = x$
   (c) Minimize $f = 2x + 3y$
   (d) Maximize $f = 2x + 3y$
   (e) Maximize $f = 3x + 2y$
   (f) Maximize $f = x + y$

5. Referring to the feasibility set in Figure 5.38:

   (a) Minimize $f = \dfrac{1}{2}x + y + 1$

(b)   Minimize $f = x$
(c)   Maximize $f = x + 2y$
(d)   Maximize $f = 2x + y$

6.   How many extreme points does a polygonal convex set with n sides in the plane have?

7.   A small dairy produces a Cheddar cheese and a Cheddar cheese spread. The production costs and monthly constraints are as follows:

| | Cost, dollars per pound | | Maximum monthly allocation, dollars |
	Cheddar	Cheddar spread	
Ingredients	0.5	0.2	1,200
Plant cost	0.5	0.5	1,500
Labor	0.2	1	2,200

How can the dairy maximize its profit from these two items if it charges its customers $4.00 per pound for the Cheddar, and $8.00 per pound for the Cheddar spread?

8.   In Exercise 7, suppose that the dairy was able to double its monthly allocation for ingredients, plant cost, and labor. To increase sales, it reduced the price of the Cheddar to $3.50 per pound and the price of the Cheddar spread to $6.00 per pound. How can the dairy maximize its profit on these two items?

6
MATRICES AND LINEAR EQUATIONS

6

Geometric methods and intuition are very important and useful tools in mathematics. They help us get insights and understanding of abstract problems and techniques and, at times, even provide effective means for finding solutions. Such was the case in Chapter 5. Sooner or later, however, geometric intuition has to give way to abstraction and more formal but efficient methods. We have reached this point now. In order to progress effectively in our study of probability and linear programming, we must study new techniques.

Matrices have many interpretations and applications. We present below only a brief introduction to the subject, just sufficient for our purposes in this text. Matrix theory belongs to the branch of algebra called *linear algebra*.

6.1 MATRIX ADDITION AND MULTIPLICATION BY SCALARS

By way of motivation, consider a system of linear equations

$$a_1x + b_1y + c_1z = d_1$$
$$a_2x + b_2y + c_2z = d_2$$
$$a_3x + b_3y + c_3z = d_3$$

in the unknowns x, y, and z. The coefficients a_i, b_i, and c_i are written in a rectangular array called a *matrix* (see Figure 6.1). It has *three rows* and *three columns*, and it is said to be a 3 by 3 matrix. The entries of a matrix are called *elements*.

Another example of a matrix is given in Figure 6.2. This matrix has *three rows* and *one column*.

In general, an m by n matrix A is a rectangular array having m rows and n columns (see Figure 6.3). Designating the elements by a_{ij} with the double index is very convenient here. The first index designates the row and the second index designates the column. Thus, a_{73} lies at the intersection of the seventh row and the third column.

$$
\begin{array}{ccc}
\text{First column} & \text{Second column} & \text{Third column}
\end{array}
$$

$$
\begin{bmatrix} a_1 & b_1 & c_1 \\ a_2 & b_2 & c_2 \\ a_3 & b_3 & c_3 \end{bmatrix}
\begin{array}{l} \text{First row} \\ \text{Second row} \\ \text{Third row} \end{array}
$$

FIGURE 6.1 *Example of a 3 by 3 matrix*

$$
\begin{bmatrix} d_1 \\ d_2 \\ d_3 \end{bmatrix}
\begin{array}{l} \text{First row} \\ \text{Second row} \\ \text{Third row} \end{array}
$$

FIGURE 6.2 *Example of a 3 by 1 matrix*

$$
\begin{array}{ccccccc}
\text{First column} & \text{Second column} & \text{Third column} & & j\text{th column} & & n\text{th column}
\end{array}
$$

$$
\begin{bmatrix}
a_{11} & a_{12} & a_{13} & \cdots & a_{1j} & \cdots & a_{1n} \\
a_{21} & a_{22} & a_{23} & \cdots & a_{2j} & \cdots & a_{2n} \\
a_{31} & a_{32} & a_{33} & \cdots & a_{3j} & \cdots & a_{3n} \\
\cdot & \cdot & \cdot & & \cdot & & \cdot \\
\cdot & \cdot & \cdot & & \cdot & & \cdot \\
\cdot & \cdot & \cdot & & \cdot & & \cdot \\
a_{i1} & a_{i2} & a_{i3} & \cdots & a_{ij} & \cdots & a_{im} \\
\cdot & \cdot & \cdot & & \cdot & & \cdot \\
\cdot & \cdot & \cdot & & \cdot & & \cdot \\
\cdot & \cdot & \cdot & & \cdot & & \cdot \\
a_{m1} & a_{m2} & a_{m3} & \cdots & a_{mj} & \cdots & a_{mn}
\end{bmatrix}
\begin{array}{l} \text{First row} \\ \text{Second row} \\ \text{Third row} \\ \\ \\ \\ i\text{th row} \\ \\ \\ \\ m\text{th row} \end{array}
$$

FIGURE 6.3 *Example of an* m *by* n *matrix*

The matrix in Figure 6.3 is often expressed in the abbreviated form

$$A = [a_{ij}]$$

Example 1

(1) $A = \begin{bmatrix} 1 & 2 \\ 3 & -1 \end{bmatrix}$ is a 2 by 2, or *square* matrix

(2) $B = \begin{bmatrix} 1 & 2 & 0 \\ 0 & -1 & -2 \end{bmatrix}$ is a 2 by 3 matrix

(3) $C = \begin{bmatrix} 1 & 0 & 0 \\ 0 & \frac{2}{3} & 0 \\ 0 & 0 & -\frac{1}{3} \end{bmatrix}$ is a 3 by 3 matrix, called a *diagonal* matrix because all entries off the diagonal are zero

Terminology

The *dimension* or *shape* of a matrix is its number of rows and columns.

> An n by n matrix is called a *square matrix*.
> An m by 1 matrix is called a *column vector*.
> A 1 by n matrix is called a *row vector*.

Now, to make matrices useful, we must be able to manipulate them. Matrix operations are defined only for matrices of compatible shapes, as explained below. We begin with the following definition.

DEFINITION 1

Two matrices $A = [a_{ij}]$ and $B = [b_{ij}]$ are *equal*, written $A = B$, if they have the same shape and if $a_{ij} = b_{ij}$ for all values of i and j.

DEFINITION 2

Consider the matrices

$$A = \begin{bmatrix} a_{11} & a_{12} & \cdots & a_{1n} \\ a_{21} & a_{22} & \cdots & a_{2n} \\ \cdot & \cdot & & \cdot \\ \cdot & \cdot & & \cdot \\ \cdot & \cdot & & \cdot \\ a_{m1} & a_{m2} & \cdots & a_{mn} \end{bmatrix} \quad \text{and} \quad B = \begin{bmatrix} b_{11} & b_{12} & \cdots & b_{1n} \\ b_{21} & b_{22} & \cdots & b_{2n} \\ \cdot & \cdot & & \cdot \\ \cdot & \cdot & & \cdot \\ \cdot & \cdot & & \cdot \\ b_{m1} & b_{m2} & \cdots & b_{mn} \end{bmatrix}$$

The *sum* matrix $A + B$ is defined by

$$A + B = \begin{bmatrix} a_{11} + b_{11} & a_{12} + b_{12} & \cdots & a_{1n} + b_{1n} \\ a_{21} + b_{21} & a_{22} + b_{22} & \cdots & a_{2n} + b_{2n} \\ \cdot & \cdot & & \cdot \\ \cdot & \cdot & & \cdot \\ a_{m1} + b_{m1} & a_{m2} + b_{m2} & \cdots & a_{mn} + b_{mn} \end{bmatrix}$$

and for any number (scalar) c,

$$cA = \begin{bmatrix} ca_{11} & ca_{12} & \cdots & ca_{1n} \\ ca_{21} & ca_{22} & \cdots & ca_{2n} \\ \cdot & \cdot & & \cdot \\ \cdot & \cdot & & \cdot \\ \cdot & \cdot & & \cdot \\ ca_{m1} & ca_{m2} & \cdots & ca_{mn} \end{bmatrix}$$

Observe that addition is defined only for matrices having the same shape. If $A = [a_{ij}]$, $B = [b_{ij}]$, and $A + B = [c_{ij}]$, then $c_{ij} = a_{ij} + b_{ij}$. Thus, $A + B$ is formed by adding the entries which occupy the same position in A and in B. The matrix cA is obtained by multiplying each element of A by c.

Example 2
Find $A + 5B$ when

$$A = \begin{bmatrix} 4 \\ -1 \\ 2 \end{bmatrix} \quad \text{and} \quad B = \begin{bmatrix} 1 \\ 2 \\ -1 \end{bmatrix}$$

SOLUTION

$$A + 5B = \begin{bmatrix} 4 \\ -1 \\ 2 \end{bmatrix} + 5\begin{bmatrix} 1 \\ 2 \\ -1 \end{bmatrix} = \begin{bmatrix} 4 \\ -1 \\ 2 \end{bmatrix} + \begin{bmatrix} 5 \\ 10 \\ -5 \end{bmatrix}$$

$$= \begin{bmatrix} 4 + 5 \\ -1 + 10 \\ 2 + (-5) \end{bmatrix} = \begin{bmatrix} 9 \\ 9 \\ -3 \end{bmatrix}$$

Example 3
Find $4A + 3B$ when

$$A = [1 \quad 0 \quad 2 \quad -1] \quad \text{and} \quad B = [4 \quad -1 \quad 0 \quad 2]$$

SOLUTION

$$\begin{aligned} 4A + 3B &= 4[1 \quad 0 \quad 2 \quad -1] + 3[4 \quad -1 \quad 0 \quad 2] \\ &= [4 \quad 0 \quad 8 \quad -4] + [12 \quad -3 \quad 0 \quad 6] \\ &= [4 + 12 \quad 0 + (-3) \quad 8 + 0 \quad -4 + 6] \\ &= [16 \quad -3 \quad 8 \quad 2] \end{aligned}$$

Example 4

Find $A + B$ and $3A$ when

$$A = \begin{bmatrix} 1 & -2 & 0 & 3 \\ 2 & 1 & 3 & -1 \\ 0 & -3 & 1 & 0 \end{bmatrix} \quad \text{and} \quad B = \begin{bmatrix} 2 & 0 & -1 & -1 \\ 5 & -1 & 2 & 0 \\ -1 & 7 & 3 & 0 \end{bmatrix}$$

SOLUTION

$$A + B = \begin{bmatrix} 1+2 & -2+0 & 0+(-1) & 3+(-1) \\ 2+5 & 1+(-1) & 3+2 & -1+0 \\ 0+(-1) & -3+7 & 1+3 & 0+0 \end{bmatrix}$$

$$= \begin{bmatrix} 3 & -2 & -1 & 2 \\ 7 & 0 & 5 & -1 \\ -1 & 4 & 4 & 0 \end{bmatrix}$$

$$3A = \begin{bmatrix} 3 & -6 & 0 & 9 \\ 6 & 3 & 9 & -3 \\ 0 & -9 & 3 & 0 \end{bmatrix}$$

Example 5

Find $3A - 2B$ when A and B are as in Example 4.

SOLUTION

By the matrix $3A - 2B$ we mean the matrix $3A + (-2)B$. Since

$$-2B = \begin{bmatrix} -4 & 0 & 2 & 2 \\ -10 & 2 & -4 & 0 \\ 2 & -14 & -6 & 0 \end{bmatrix}$$

we find that

$$3A - 2B = \begin{bmatrix} 3+(-4) & -6+0 & 0+2 & 9+2 \\ 6+(-10) & 3+2 & 9+(-4) & -3+0 \\ 0+2 & -9+(-14) & 3+(-6) & 0+0 \end{bmatrix}$$

$$= \begin{bmatrix} -1 & -6 & 2 & 11 \\ -4 & 5 & 5 & -3 \\ 2 & -23 & -3 & 0 \end{bmatrix}$$

We conclude this section with a list of basic properties of matrix addition and multiplication by numbers. The symbol 0 is used both to designate the number zero as well as the matrix whose entries are all zero. The meaning of 0 will always be clear from the context.

Properties of matrix addition and multiplication by numbers

Let the matrices A, B, and C have the same shape, and let r and s be any numbers. Then

(1) $A + B = B + A$
(2) $A + (B + C) = (A + B) + C$
(3) $A + 0 = A$
(4) $A + (-1)A = 0$
(5) $r(sA) = (rs)A$
(6) $r(A + B) = rA + rB$
(7) $(r + s)A = rA + sA$
(8) $1 \cdot A = A$

We thus see that there is great flexibility in matrix manipulations. The multiplication of matrices, which is a more complicated operation, is discussed in Section 6.2.

EXERCISES

Compute the matrices in Exercises 1–11.

1. $2[1 \quad -1 \quad 0] + 4[-1 \quad 0 \quad 1]$

2. $-7[1 \quad 1 \quad 1] + 7[-1 \quad -1 \quad -1] - 2[1 \quad -1 \quad 0]$

3. $[\frac{1}{6} \quad 0 \quad \frac{5}{6}] + \frac{1}{3}[1 \quad -2 \quad 0]$

4. $\begin{bmatrix} 2 \\ 4 \end{bmatrix} - \frac{1}{2}\begin{bmatrix} 3 \\ 6 \end{bmatrix} + \frac{1}{2}\begin{bmatrix} -1 \\ 1 \end{bmatrix}$

5. $\begin{bmatrix} 1 \\ 0 \\ -1 \end{bmatrix} - \begin{bmatrix} -1 \\ 0 \\ 1 \end{bmatrix} + 3\begin{bmatrix} 1 \\ 5 \\ 2 \end{bmatrix}$

6. $\begin{bmatrix} 1 & 0 \\ 0 & -1 \end{bmatrix} + \begin{bmatrix} 0 & 1 \\ 1 & 0 \end{bmatrix}$

7. $\begin{bmatrix} 1 & 0 \\ 1 & 0 \end{bmatrix} + \frac{1}{2}\begin{bmatrix} 0 & 2 \\ 0 & 2 \end{bmatrix}$

8. $\begin{bmatrix} \frac{1}{2} & -\frac{1}{2} \\ -\frac{1}{3} & \frac{1}{3} \end{bmatrix} - \begin{bmatrix} \frac{1}{4} & -\frac{1}{4} \\ -\frac{1}{9} & \frac{1}{9} \end{bmatrix}$

9. $\begin{bmatrix} 2 & -1 \\ 1 & -2 \\ 3 & 2 \end{bmatrix} + 2 \begin{bmatrix} 4 & -2 \\ \frac{1}{2} & 1 \\ 1 & -\frac{1}{2} \end{bmatrix}$

10. $\begin{bmatrix} 2 & 0 & 0 \\ 0 & -2 & 0 \\ 0 & 0 & 2 \end{bmatrix} + \frac{1}{3} \begin{bmatrix} 3 & 0 & 0 \\ 0 & -3 & 0 \\ 0 & 0 & 6 \end{bmatrix}$

11. $\begin{bmatrix} 1 & 2 & 0 & 0 \\ -2 & 1 & 0 & 0 \\ 0 & 0 & 4 & 1 \\ 0 & 0 & 1 & -4 \end{bmatrix} + \begin{bmatrix} 6 & 3 & 0 & 0 \\ -1 & 1 & 0 & 0 \\ 0 & 0 & 2 & -1 \\ 0 & 0 & -1 & 2 \end{bmatrix}$

12. Let

$$A = \begin{bmatrix} -2 & 1 \\ 1 & -3 \end{bmatrix} \qquad B = \begin{bmatrix} -2 & 1 & 1 \\ 1 & 2 & -3 \end{bmatrix}$$

$$C = \begin{bmatrix} 0 & 0 & 1 \\ 0 & 1 & 0 \\ 1 & 0 & 0 \end{bmatrix} \qquad D = \begin{bmatrix} 0 & 1 \\ -1 & 0 \\ 0 & 1 \end{bmatrix}$$

$$E = \begin{bmatrix} \frac{1}{2} & \frac{1}{3} \\ \frac{1}{3} & 0 \\ 0 & 0 \end{bmatrix} \qquad F = \begin{bmatrix} \frac{1}{3} & -\frac{1}{3} & \frac{1}{3} \\ -\frac{1}{2} & \frac{1}{2} & -\frac{1}{2} \\ 1 & 0 & 0 \end{bmatrix}$$

Compute the following matrices if they are defined.

(a) $A + B$
(b) $\frac{1}{2}C + F$
(c) $E - D$
(d) $B + D$
(e) $C - 2E$
(f) $3E + C$
(g) $C - F$

13. Referring to Exercise 12: Find in each case below a matrix X satisfying the given equation.

(a) $A + X = \frac{1}{2}A$
(b) $D + X = E$
(c) $F + X = 2C$
(d) $D + X = 0$
(e) $2F + X = 3F$
(f) $A + X = 0$

6.2 MATRIX MULTIPLICATION

We introduce the concept of matrix multiplication through a formal definition of product of a row vector with a column vector.

DEFINITION 1

Consider a 1 by n row vector R and an n by 1 column vector C:

$$R = [r_1 \quad r_2 \quad r_3 \quad \cdots \quad r_n] \quad \text{and} \quad C = \begin{bmatrix} c_1 \\ c_2 \\ c_3 \\ \cdot \\ \cdot \\ \cdot \\ c_n \end{bmatrix}$$

The *product* $R \cdot C$ is defined as

$$R \cdot C = r_1 c_1 + r_2 c_2 + r_3 c_3 + \cdots + r_n c_n$$

Observe that for the product to be defined, the two vectors must have the same number of entries. Of the various interpretations this product can be given, we consider one in Example 2 below.

Example 1

$$[1 \quad 1 \quad 1] \cdot \begin{bmatrix} 2 \\ 3 \\ -2 \end{bmatrix} = 1 \cdot 2 + 1 \cdot 3 + 1 \cdot (-2) = 2 + 3 - 2 = 3$$

$$[0 \quad 0 \quad 0] \cdot \begin{bmatrix} 2 \\ 3 \\ -2 \end{bmatrix} = 0 \cdot 2 + 0 \cdot 3 + 0 \cdot (-2) = 0$$

$$[1 \quad 3] \cdot \begin{bmatrix} -3 \\ 1 \end{bmatrix} = 1 \cdot (-3) + 3 \cdot 1 = -3 + 3 = 0$$

$$[2 \quad 2 \quad -5] \cdot \begin{bmatrix} 2 \\ 3 \\ 2 \end{bmatrix} = 2 \cdot 2 + 2 \cdot 3 + (-5) \cdot 2 = 4 + 6 - 10 = 0$$

We note that the product of two vectors may be zero even when neither of the vectors is the zero vector.

Example 2

In the manufacture of a small electric appliance, a firm has the following average labor expenses:

Item	Mechanical assembly	Electrical assembly	Electrical wiring	Final assembly
Hourly wage in dollars	3.6	3.8	4.5	3.6
Average number of hours per unit	2	2.2	3	1.5

Let us express this information in a "wage" vector and an "hour" vector:

$$W = [3.6 \quad 3.8 \quad 4.5 \quad 3.6] \qquad H = \begin{bmatrix} 2 \\ 2.2 \\ 3 \\ 1.5 \end{bmatrix}$$

The average labor cost per unit can now be given as a product of these vectors. By Definition 1 we have

$$W \cdot H = [3.6 \quad 3.8 \quad 4.5 \quad 3.6] \cdot \begin{bmatrix} 2 \\ 2.2 \\ 3 \\ 1.5 \end{bmatrix}$$

$$= 3.6 \times 2 + 3.8 \times 2.2 + 4.5 \times 3 + 3.6 \times 1.5$$

$$= 34.46 \text{ dollars}$$

Observe that the product of two vectors is a *number*. A number can, of course, be regarded as a 1 by 1 matrix.

To define the product of matrices in general, we now observe that a matrix may be regarded as being made up of row vectors, or as being made up of column vectors. With this in mind, the product we attempt to define must be consistent with Definition 1. Thus, if $A \cdot B = C$ for some matrices A, B, and C, then the entry in the (i, j)-position of C must be the product of the ith row of A with the jth column of B (see Figure 6.4). This in turn implies that the number of *columns* of A must equal the number of *rows* of B.

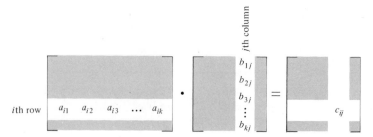

FIGURE 6.4

Example 3
Consider the matrices

$$A = \begin{bmatrix} 1 & 0 & -1 \\ 2 & 4 & 3 \end{bmatrix} \quad \text{and} \quad B = \begin{bmatrix} 5 & 8 \\ 1 & -1 \\ 5 & -4 \end{bmatrix}$$

The product of A and B is the following 2 by 2 matrix:

$$A \cdot B = \begin{bmatrix} \text{1st row of } A \times \text{1st column of } B & \text{1st row of } A \times \text{2d column of } B \\ \text{2d row of } A \times \text{1st column of } B & \text{2d row of } A \times \text{2d column of } B \end{bmatrix}$$

When we carry this out, we get

$$A \cdot B = \begin{bmatrix} 1 & 0 & -1 \\ 2 & 4 & 3 \end{bmatrix} \cdot \begin{bmatrix} 5 & 8 \\ 1 & -1 \\ 5 & -4 \end{bmatrix}$$

$$= \begin{bmatrix} 1 \cdot 5 + 0 \cdot 1 + (-1) \cdot 5 & 1 \cdot 8 + 0 \cdot (-1) + (-1) \cdot (-4) \\ 2 \cdot 5 + 4 \cdot 1 + 3 \cdot 5 & 2 \cdot 8 + 4 \cdot (-1) + 3 \cdot (-4) \end{bmatrix}$$

$$= \begin{bmatrix} 0 & 12 \\ 29 & 0 \end{bmatrix}$$

DEFINITION 2
Let $A = [a_{ij}]$ be an m by n matrix, and let $B = [b_{jk}]$ be an n by p matrix. Then the product $C = A \cdot B$ is the m by p matrix $C = [c_{ik}]$ such that

$$c_{ik} = a_{i1}b_{1k} + a_{i2}b_{2k} + a_{i3}b_{3k} + \cdots + a_{in}b_{nk}$$

for $i = 1, 2, \ldots, m$, and $k = 1, 2, \ldots, p$.

Example 4

Compute $A \cdot B$ and $B \cdot A$ when

$$A = \begin{bmatrix} 2 & 2 \\ 1 & 0 \end{bmatrix} \quad \text{and} \quad B = \begin{bmatrix} 0 & 3 \\ 1 & 3 \end{bmatrix}$$

SOLUTION

$$A \cdot B = \begin{bmatrix} 2 & 2 \\ 1 & 0 \end{bmatrix} \cdot \begin{bmatrix} 0 & 3 \\ 1 & 3 \end{bmatrix} = \begin{bmatrix} 2 \cdot 0 + 2 \cdot 1 & 2 \cdot 3 + 2 \cdot 3 \\ 1 \cdot 0 + 0 \cdot 1 & 1 \cdot 3 + 0 \cdot 3 \end{bmatrix}$$

$$= \begin{bmatrix} 2 & 12 \\ 0 & 3 \end{bmatrix}$$

$$B \cdot A = \begin{bmatrix} 0 & 3 \\ 1 & 3 \end{bmatrix} \cdot \begin{bmatrix} 2 & 2 \\ 1 & 0 \end{bmatrix} = \begin{bmatrix} 0 \cdot 2 + 3 \cdot 1 & 0 \cdot 2 + 3 \cdot 0 \\ 1 \cdot 2 + 3 \cdot 1 & 1 \cdot 2 + 3 \cdot 0 \end{bmatrix}$$

$$= \begin{bmatrix} 3 & 0 \\ 5 & 2 \end{bmatrix}$$

We observe that, unlike multiplication of numbers, the multiplication of matrices is, in general, not commutative: That is, the order in which the matrices are multiplied is important, and in general

$$A \cdot B \neq B \cdot A$$

Example 5

Calculate $A^2 = A \cdot A$ when

$$A = \begin{bmatrix} 1 & 2 \\ 0 & 2 \end{bmatrix}$$

SOLUTION

$$A \cdot A = \begin{bmatrix} 1 & 2 \\ 0 & 2 \end{bmatrix} \cdot \begin{bmatrix} 1 & 2 \\ 0 & 2 \end{bmatrix} = \begin{bmatrix} 1 \cdot 1 + 2 \cdot 0 & 1 \cdot 2 + 2 \cdot 2 \\ 0 \cdot 1 + 2 \cdot 0 & 0 \cdot 2 + 2 \cdot 2 \end{bmatrix}$$

$$= \begin{bmatrix} 1 & 6 \\ 0 & 4 \end{bmatrix}$$

Example 6

Calculate $(A + 4B) \cdot B$ when

$$A = \begin{bmatrix} 1 & 2 \\ 2 & 3 \end{bmatrix} \quad \text{and} \quad B = \begin{bmatrix} 1 & 1 \\ 1 & 0 \end{bmatrix}$$

SOLUTION

$$(A + 4B) \cdot B = \left(\begin{bmatrix} 1 & 2 \\ 2 & 3 \end{bmatrix} + 4 \begin{bmatrix} 1 & 1 \\ 1 & 0 \end{bmatrix} \right) \cdot \begin{bmatrix} 1 & 1 \\ 1 & 0 \end{bmatrix}$$

$$= \left(\begin{bmatrix} 1 & 2 \\ 2 & 3 \end{bmatrix} + \begin{bmatrix} 4 & 4 \\ 4 & 0 \end{bmatrix} \right) \cdot \begin{bmatrix} 1 & 1 \\ 1 & 0 \end{bmatrix}$$

$$= \begin{bmatrix} 5 & 6 \\ 6 & 3 \end{bmatrix} \cdot \begin{bmatrix} 1 & 1 \\ 1 & 0 \end{bmatrix}$$

$$= \begin{bmatrix} 5 + 6 & 5 \\ 6 + 3 & 6 \end{bmatrix} = \begin{bmatrix} 11 & 5 \\ 9 & 6 \end{bmatrix}$$

Example 7

Calculate $A \cdot B + 4B^2$ when A and B are as in Example 6.

SOLUTION

$$A \cdot B = \begin{bmatrix} 1 & 2 \\ 2 & 3 \end{bmatrix} \cdot \begin{bmatrix} 1 & 1 \\ 1 & 0 \end{bmatrix} = \begin{bmatrix} 1 + 2 & 1 \\ 2 + 3 & 2 \end{bmatrix} = \begin{bmatrix} 3 & 1 \\ 5 & 2 \end{bmatrix}$$

$$4B^2 = 4 \begin{bmatrix} 1 & 1 \\ 1 & 0 \end{bmatrix} \cdot \begin{bmatrix} 1 & 1 \\ 1 & 0 \end{bmatrix} = 4 \begin{bmatrix} 1 + 1 & 1 \\ 1 & 1 \end{bmatrix}$$

$$= 4 \begin{bmatrix} 2 & 1 \\ 1 & 1 \end{bmatrix} = \begin{bmatrix} 8 & 4 \\ 4 & 4 \end{bmatrix}$$

Hence, we find that

$$A \cdot B + 4B^2 = \begin{bmatrix} 3 & 1 \\ 5 & 2 \end{bmatrix} + \begin{bmatrix} 8 & 4 \\ 4 & 4 \end{bmatrix} = \begin{bmatrix} 3 + 8 & 1 + 4 \\ 5 + 4 & 2 + 4 \end{bmatrix}$$

$$= \begin{bmatrix} 11 & 5 \\ 9 & 6 \end{bmatrix}$$

Comparing Examples 6 and 7 we see that

$$(A + 4B) \cdot B = A \cdot B + 4B \cdot B$$

that is, matrix multiplication is *distributive* in this case as well as in general, and the following properties hold whenever the products are defined.

Properties of matrix multiplication

If the products are defined, then

(1) $A \cdot (B + C) = A \cdot B + A \cdot C$
(2) $(A + B) \cdot C = A \cdot C + B \cdot C$
(3) $A \cdot (B \cdot C) = (A \cdot B) \cdot C$
(4) $(sA) \cdot B = A \cdot (sB) = s(A \cdot B)$ for any scalar s

Consider now a system of linear equations

$$a_{11}x + a_{12}y = b_1$$
$$a_{21}x + a_{22}y = b_2$$

With the notation

$$A = \begin{bmatrix} a_{11} & a_{12} \\ a_{21} & a_{22} \end{bmatrix} \quad X = \begin{bmatrix} x \\ y \end{bmatrix} \quad \text{and} \quad B = \begin{bmatrix} b_1 \\ b_2 \end{bmatrix}$$

this system can be written as the single equation

$$A \cdot X = B$$

From Section 5.3 we know how to find the vector X when it exists.

EXERCISES

1. Find the products below when

$$A = \begin{bmatrix} 1 & 2 \\ 3 & 4 \end{bmatrix} \quad B = \begin{bmatrix} 1 & 3 \\ 2 & 4 \end{bmatrix} \quad I = \begin{bmatrix} 1 & 0 \\ 0 & 1 \end{bmatrix} \quad \text{and}$$

$$J = \begin{bmatrix} 0 & 1 \\ 1 & 0 \end{bmatrix}$$

(a) $A \cdot B$
(b) $B \cdot A$
(c) $A \cdot I$
(d) $I \cdot A$
(e) $A \cdot J$
(f) $J \cdot J$
(g) $I \cdot I$
(h) $(I + J) \cdot (I + J)$
(i) $(A + I) \cdot J$
(j) $(2A + 3B) \cdot A$
(k) $(2I + 3J) \cdot J$

Calculate the products in Exercises 2–12.

2.
$$\begin{bmatrix} 0 & 0 & -1 \\ 0 & -1 & 0 \\ -1 & 0 & 0 \end{bmatrix} \cdot \begin{bmatrix} 1 & 0 & 0 \\ 0 & 2 & 0 \\ 0 & 0 & 3 \end{bmatrix}$$

3.
$$\begin{bmatrix} 1 & 0 & -1 \\ 0 & -1 & 0 \end{bmatrix} \cdot \begin{bmatrix} 2 & -3 \\ 3 & 0 \\ 4 & 2 \end{bmatrix}$$

4.
$$\begin{bmatrix} 2 & -3 \\ 3 & 0 \\ 4 & 2 \end{bmatrix} \cdot \begin{bmatrix} 0 & 2 \\ 3 & 0 \end{bmatrix}$$

5.
$$\begin{bmatrix} 1 \\ 0 \\ 2 \end{bmatrix} \cdot [2 \quad 3 \quad 4]$$

6.
$$\begin{bmatrix} 1 \\ -1 \\ 1 \end{bmatrix} \cdot [1 \quad 0 \quad 2]$$

7.
$$\begin{bmatrix} \frac{1}{2} & \frac{1}{3} \\ -\frac{1}{3} & \frac{1}{2} \end{bmatrix} \cdot \begin{bmatrix} 1 & 0 & 2 \\ 2 & 1 & 3 \end{bmatrix}$$

8.
$$\begin{bmatrix} 1 & -3 & -2 \\ -2 & 1 & 3 \\ 3 & -2 & 1 \end{bmatrix} \cdot \begin{bmatrix} 1 & 3 & -2 \\ -2 & 1 & 3 \\ 3 & -2 & 1 \end{bmatrix}$$

9.
$$\begin{bmatrix} 1 & 0 & 0 \\ 0 & 2 & 0 \\ 0 & 0 & 3 \end{bmatrix} \cdot \begin{bmatrix} 1 & 0 & 0 \\ 0 & 2 & 0 \\ 0 & 0 & 3 \end{bmatrix}$$

10.
$$\begin{bmatrix} 1 & -2 & 0 \\ 0 & 1 & 0 \\ 0 & 1 & 1 \end{bmatrix} \cdot \begin{bmatrix} 1 & 0 & 1 \\ 0 & 0 & 0 \\ 1 & 0 & 1 \end{bmatrix}$$

11.
$$\begin{bmatrix} 0 & 0 & 0 \\ 2 & 0 & 0 \\ 1 & 3 & 0 \end{bmatrix} \cdot \begin{bmatrix} 0 & 0 & 0 \\ 1 & 0 & 0 \\ 1 & 1 & 0 \end{bmatrix}$$

12.
$$\begin{bmatrix} 0 & 2 & 1 \\ 0 & 0 & 3 \\ 0 & 0 & 0 \end{bmatrix} \cdot \begin{bmatrix} 0 & 1 & 1 \\ 0 & 0 & 1 \\ 0 & 0 & 0 \end{bmatrix}$$

In Exercises 13–18 find the vector $X = \begin{bmatrix} x \\ y \end{bmatrix}$ which is a solution of the given equation.

Hint:

Consult Section 5.3.

13. $\begin{bmatrix} 2 & 1 \\ -1 & 2 \end{bmatrix} \cdot \begin{bmatrix} x \\ y \end{bmatrix} = \begin{bmatrix} 0 \\ 1 \end{bmatrix}$

14. $\begin{bmatrix} 1 & 2 \\ 0 & 1 \end{bmatrix} \cdot \begin{bmatrix} x \\ y \end{bmatrix} = \begin{bmatrix} 1 \\ -1 \end{bmatrix}$

15. $\begin{bmatrix} 1 & 2 \\ -2 & 3 \end{bmatrix} \cdot \begin{bmatrix} x \\ y \end{bmatrix} = \begin{bmatrix} 1 \\ 0 \end{bmatrix}$

16. $\begin{bmatrix} 1 & 3 \\ -1 & 2 \end{bmatrix} \cdot \begin{bmatrix} x \\ y \end{bmatrix} = \begin{bmatrix} 6 \\ -1 \end{bmatrix}$

$\begin{bmatrix} \frac{1}{2} & -1 \\ 2 & -1 \end{bmatrix} \cdot \begin{bmatrix} x \\ y \end{bmatrix} = \begin{bmatrix} 1 \\ 1 \end{bmatrix}$

18. $\begin{bmatrix} -4 & 1 \\ 1 & 1 \end{bmatrix} \cdot \begin{bmatrix} x \\ y \end{bmatrix} = \begin{bmatrix} 6 \\ 9 \end{bmatrix}$

In Exercises 19–22 find the squares of the given matrices ($A^2 = A \cdot A$).

19. $\begin{bmatrix} 1 & 1 & 0 \\ 0 & 0 & 1 \\ 1 & 0 & 0 \end{bmatrix}$

20. $\begin{bmatrix} 1 & 0 & 0 & 0 \\ 0 & 0 & 1 & 1 \\ 0 & 1 & 0 & 0 \\ 0 & 0 & 0 & 1 \end{bmatrix}$

21. $\begin{bmatrix} 1 & 2 & 0 & 0 \\ 0 & 0 & 1 & 1 \\ 0 & 0 & 0 & 0 \\ 1 & 0 & 0 & 0 \end{bmatrix}$

22. $\begin{bmatrix} 1 & 0 & 1 & 1 \\ 0 & 1 & 1 & 1 \\ 1 & 1 & 0 & 1 \\ 1 & 1 & 1 & 0 \end{bmatrix}$

6.3 SYSTEMS OF LINEAR EQUATIONS IN MORE THAN TWO UNKNOWNS

In Section 5.3 we discussed the solutions of systems of two linear equations in two unknowns; in Section 6.2 we saw how to express such a system in matrix notation. We shall now show how matrices are used in solving general systems of linear equations. The method discussed below is known as the *pivotal reduction* method, or *Gaussian elimination*.

Example 1
Solve the system of linear equations

$$2x + y - 5z = 2$$
$$6x - 4y + 3z = 3 \tag{1}$$
$$5x + \frac{1}{2}y - \frac{11}{2}z = 6$$

in the unknowns x, y, and z.

SOLUTION 1
To illustrate the method, we first solve this problem without matrix notation. The basic idea is to eliminate x from the second equation and x and y from the third equation. We shall then be able to find the solution if it exists.

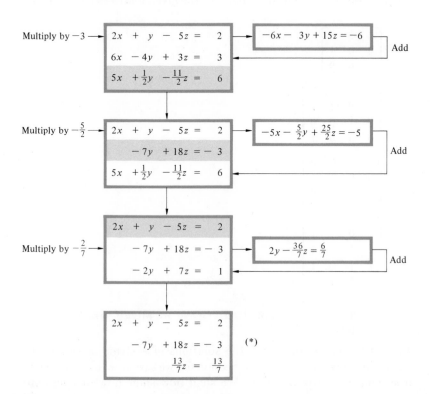

In the reduction presented schematically above, x is used as a pivot in the sense that it remains fixed throughout the process of reduction. The shaded rows are those not used in that particular step.

The last system (∗), which is equivalent to the original system, is now easy to solve: The last equation gives $z = 1$; substituting this into the second equation and solving for y gives

$$y = \frac{1}{7}(18z + 3) = \frac{1}{7}(18 \cdot 1 + 3) = 3$$

Substituting for y and z in the first equation and solving for x gives

$$x = \frac{1}{2}(2 - y + 5z) = \frac{1}{2}(2 - 3 + 5) = 2$$

The solution is thus $x = 2$, $y = 3$, $z = 1$.

SOLUTION 2

In matrix notation, the system in (1) can be written as

$$\begin{bmatrix} 2 & 1 & -5 \\ 6 & -4 & 3 \\ 5 & \frac{1}{2} & -\frac{11}{2} \end{bmatrix} \cdot \begin{bmatrix} x \\ y \\ z \end{bmatrix} = \begin{bmatrix} 2 \\ 3 \\ 6 \end{bmatrix}$$

We combine the coefficient matrix and the column vector of constants into an *augmented matrix*

$$\left[\begin{array}{ccc|c} 2 & 1 & -5 & 2 \\ 6 & -4 & 3 & 3 \\ 5 & \frac{1}{2} & -\frac{11}{2} & 6 \end{array}\right]$$

For convenience, the column vector of constants has been separated from the coefficient matrix by a vertical line. The method is now presented schematically on page 262.

We could now use the last matrix (∗∗) to write the system in the form (∗) and then proceed as above. We note, however, that the solution of (1) can be expressed as

$$1 \cdot x + 0 \cdot y + 0 \cdot z = 2$$
$$0 \cdot x + 1 \cdot y + 0 \cdot z = 3$$
$$0 \cdot x + 0 \cdot y + 1 \cdot z = 1$$

Multiply by $-3 \longrightarrow$
$$\left[\begin{array}{ccc|c} 2 & 1 & -5 & 2 \\ 6 & -4 & 3 & 3 \\ 5 & \frac{1}{2} & -\frac{11}{2} & 6 \end{array}\right] \longleftarrow\!-\!\left[-6 \quad -3 \quad 15 \mid -6\right]$$
Add

$\downarrow$

Multiply by $-\frac{5}{2} \longrightarrow$
$$\left[\begin{array}{ccc|c} 2 & 1 & -5 & 2 \\ 0 & -7 & 18 & -3 \\ 5 & \frac{1}{2} & -\frac{11}{2} & 6 \end{array}\right] \longleftarrow\!-\!\left[-5 \quad -\frac{5}{2} \quad \frac{25}{2} \mid -5\right]$$
Add

$\downarrow$

Multiply by $-\frac{2}{7} \longrightarrow$
$$\left[\begin{array}{ccc|c} 2 & 1 & -5 & 2 \\ 0 & -7 & 18 & -3 \\ 0 & -2 & 7 & 1 \end{array}\right] \longleftarrow\!-\!\left[0 \quad 2 \quad -\frac{36}{7} \mid \frac{6}{7}\right]$$
Add

$\downarrow$

$$\left[\begin{array}{ccc|c} 2 & 1 & -5 & 2 \\ 0 & -7 & 18 & -3 \\ 0 & 0 & \frac{13}{7} & \frac{13}{7} \end{array}\right] \quad (**)$$

In matrix notation this becomes

$$\begin{bmatrix} 1 & 0 & 0 \\ 0 & 1 & 0 \\ 0 & 0 & 1 \end{bmatrix} \cdot \begin{bmatrix} x \\ y \\ z \end{bmatrix} = \begin{bmatrix} 2 \\ 3 \\ 1 \end{bmatrix}$$

We shall thus attempt to reduce (**) to the form

$$\left[\begin{array}{ccc|c} 1 & 0 & 0 & 2 \\ 0 & 1 & 0 & 3 \\ 0 & 0 & 1 & 1 \end{array}\right]$$

This is done with a "backward" reduction as carried out schematically on page 263.

We have thus arrived at the solution of the system (1).

Remark

A moment's reflection will show that any one of the unknowns could be used as a pivot in the preceding reduction. Also, the equations can be rearranged in any order without changing the answer. Practice will teach you many shortcuts and computation-saving ideas.

In the problems below we shall be more economical in listing calculations, but the reader should be able to follow the steps and to verify the solution.

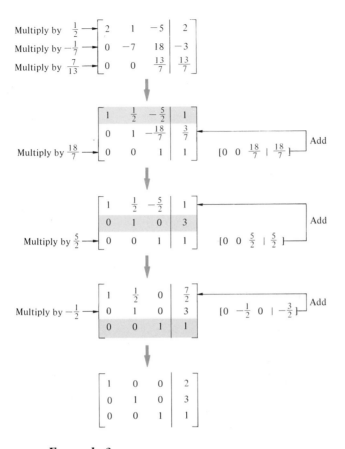

Example 2

Solve the system of linear equations

$$x + y - z = 1$$
$$3x + 2y - z = 0$$
$$4x - y + 2z = -1$$

SOLUTION

In matrix notation this system can be written as

$$\begin{bmatrix} 1 & 1 & -1 \\ 3 & 2 & -1 \\ 4 & -1 & 2 \end{bmatrix} \cdot \begin{bmatrix} x \\ y \\ z \end{bmatrix} = \begin{bmatrix} 1 \\ 0 \\ -1 \end{bmatrix}$$

We write down the augmented matrix of the system, and proceed as in Solution 2 of Example 1.

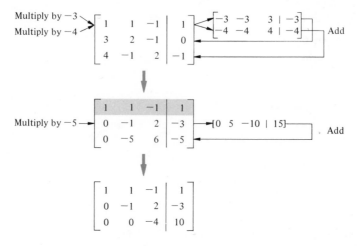

We now multiply the second row by -1 and the last row by $-\frac{1}{4}$, and then do the "backward" reduction illustrated in Example 1.

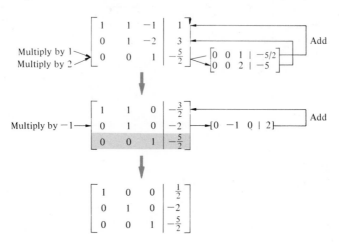

The solution is therefore

$$x = \frac{1}{2} \qquad y = -2 \qquad z = -\frac{5}{2}$$

Example 3
Solve the system of equations

$$11x + 5y + 2z = 6$$
$$5x + 5y + 2z = 4$$
$$3x + 6y - z = -3$$

SOLUTION

The augmented matrix of this system is

$$\left[\begin{array}{ccc|c} 11 & 5 & 2 & 6 \\ 5 & 5 & 2 & 4 \\ 3 & 6 & -1 & -3 \end{array}\right]$$

An inspection of this matrix shows that a change of procedure is called for this time: Instead of using 11 as a pivot to get two zeros in the first column, it is more reasonable to start with two zeros in the second or third column.

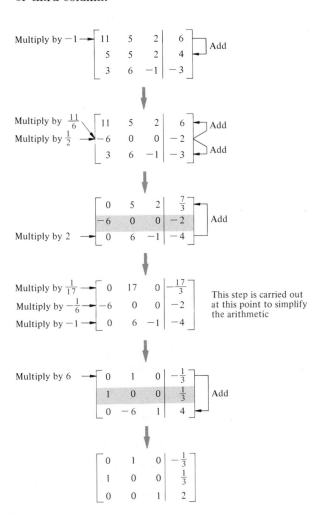

We thus arrived at the system

$$\begin{bmatrix} 0 & 1 & 0 \\ 1 & 0 & 0 \\ 0 & 0 & 1 \end{bmatrix} \cdot \begin{bmatrix} x \\ y \\ z \end{bmatrix} = \begin{bmatrix} -\frac{1}{3} \\ \frac{1}{3} \\ 2 \end{bmatrix}$$

and we find that the solution is $x = \frac{1}{3}$, $y = -\frac{1}{3}$, and $z = 2$. We observe that we have found the solution whenever the coefficient matrix reduced to a matrix having a *single 1 in every row and every column and zeros elsewhere*. When a system of linear equations is *inconsistent* (has no solution), then this will show up in the reduction.

In concluding this section, we apply the pivotal reduction to a system of four linear equations in four unknowns. The method applies to a general system of n equations in n unknowns as well. Systems where the number of equations does not equal the number of unknowns are discussed in Section 6.5.

Example 4
Solve the system of equations

$$\begin{array}{rcl} x + 3y \phantom{{}+{}} + t &=& 3 \\ 4y - z + 2t &=& -1 \\ 3x - 2y \phantom{{}+{}} + 3t &=& -2 \\ 6x + 5y - 2z + 5t &=& 5 \end{array}$$

SOLUTION

In matrix notation this system becomes

$$\begin{bmatrix} 1 & 3 & 0 & 1 \\ 0 & 4 & -1 & 2 \\ 3 & -2 & 0 & 3 \\ 6 & 5 & -2 & 5 \end{bmatrix} \cdot \begin{bmatrix} x \\ y \\ z \\ t \end{bmatrix} = \begin{bmatrix} 3 \\ -1 \\ -2 \\ 5 \end{bmatrix}$$

We follow the procedure of Examples 1–3, beginning with the augmented matrix (see page 267). As before, a shaded row indicates that this row is not involved in the given step.

From the last matrix on page 267, we find that the solution is $x = 2$, $y = 1$, $z = 1$, $t = -2$.

We can summarize the method used here for solving systems of linear equations, as follows: Rows of the augmented matrix are successively multiplied by suitable numbers and then added to other rows. The objective is to get as many zeros as possible. The process

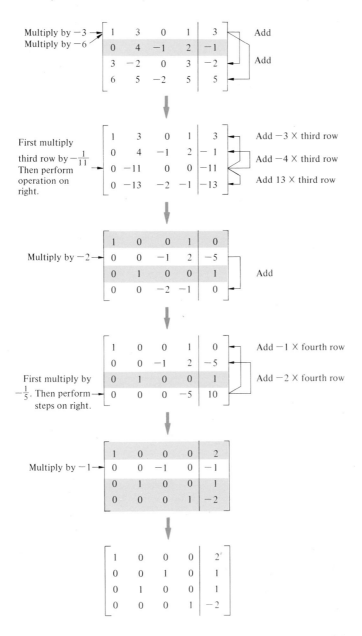

terminates when each row and each column of the coefficient matrix has only *one* nonzero entry. Finally, the coefficient matrix is reduced to a matrix having only the single nonzero entry 1 in each row and each column.

EXERCISES

Use the matrix method of this section to solve the systems of linear
equations in Exercises 1–8.

1. $x + y\ \ \ \ \ \ = -1$
   $\quad\ \ y + z =\ \ \ 1$
   $x + y + z =\ \ \ 0$

2. $\ \ x + 2y +\ z =\ \ \ 1$
   $3x +\ \ y + 2z =\ \ \ 2$
   $\ \ x +\ \ y +\ \ z = -1$

3. $-\ x +\ y -\ z = 0$
   $-2x + 7y + 2z = 1$
   $\quad x +\quad\quad\ z = 0$

4. $\ \ x - 2y + 3z = 1$
   $2x +\ \ y - 3z = 0$
   $-x +\ \ y + 2z = 0$

5. $\ \ x - 2y + 3z = 0$
   $2x +\ \ y - 3z = 1$
   $-x +\ \ y + 2z = 0$

6. $2x - y + 3z =\ \ \ 3$
   $\ \ x - y - 2z = -1$
   $\ \ x + y -\ \ z =\ \ \ 1$

7. $\ \ x + 2y - z = 6$
   $3x -\ \ y + z = 1$
   $5x + 3y + z = 1$

8. $\ \ x +\ \ y + z +\ \ t = 1$
   $\ \ x -\ \ y - z\ \ \ \ \ \ = 0$
   $-x +\ \ y + z + 2t = 0$
   $\quad\ 3y + z +\ \ t = 0$

6.4 THE INVERSE OF A SQUARE MATRIX

Consider any n by n matrix

$$A = \begin{bmatrix} a_{11} & a_{12} & \cdots & a_{1n} \\ a_{21} & a_{22} & \cdots & a_{2n} \\ \cdot & \cdot & & \cdot \\ \cdot & \cdot & & \cdot \\ \cdot & \cdot & & \cdot \\ a_{n1} & a_{n2} & \cdots & a_{nn} \end{bmatrix} \tag{2}$$

and the n by n matrix

$$I = \begin{bmatrix} 1 & 0 & 0 & \cdots & 0 \\ 0 & 1 & 0 & \cdots & 0 \\ 0 & 0 & 1 & \cdots & 0 \\ \cdot & \cdot & \cdot & & \cdot \\ \cdot & \cdot & \cdot & & \cdot \\ \cdot & \cdot & \cdot & & \cdot \\ 0 & 0 & 0 & \cdots & 1 \end{bmatrix}$$

The product of the ith row of A and the jth column of I is

$$[a_{i1} \quad a_{i2} \quad \cdots \quad a_{ij} \quad \cdots \quad a_{in}] \cdot \begin{bmatrix} 0 \\ 0 \\ \cdot \\ \cdot \\ \cdot \\ 0 \\ 1 \\ \cdot \\ \cdot \\ \cdot \\ 0 \end{bmatrix}$$

$$= a_{i1} \cdot 0 + a_{i2} \cdot 0 + \cdots + a_{ij} \cdot 1 + \cdots + a_{in} \cdot 0 = a_{ij}$$

Since this is true for $i = 1, 2, \ldots, n$ and $j = 1, 2, \ldots, n$, we conclude that $A \cdot I = A$. Likewise we show that $I \cdot A = A$. The matrix I is called the *identity* matrix. As you see, it has the same multiplicative property that 1 has in the domain of real numbers.

In many problems it is necessary to find for a given matrix A a matrix B such that $A \cdot B = I$. This matrix B, if it exists, is unique and is called the inverse of A.

DEFINITION 1

The *inverse* of a square matrix A is a matrix A^{-1} such that

$$A \cdot A^{-1} = A^{-1} \cdot A = I$$

The matrix A is called *invertible* when the inverse exists; *noninvertible* otherwise. In the literature, an invertible matrix is often called *nonsingular*; a noninvertible matrix is called *singular*.

Example 1
The matrices

$$A = \begin{bmatrix} 0 & 2 \\ -2 & 0 \end{bmatrix} \quad \text{and} \quad A^{-1} = \begin{bmatrix} 0 & -\frac{1}{2} \\ \frac{1}{2} & 0 \end{bmatrix}$$

are inverses of one another, since

$$A \cdot A^{-1} = \begin{bmatrix} 0 & 2 \\ -2 & 0 \end{bmatrix}\begin{bmatrix} 0 & -\frac{1}{2} \\ \frac{1}{2} & 0 \end{bmatrix}$$

$$= \begin{bmatrix} 0 \times 0 + 2 \times \frac{1}{2} & 0 \times (-\frac{1}{2}) + 2 \times 0 \\ -2 \times 0 + 0 \times \frac{1}{2} & -2 \times (-\frac{1}{2}) + 0 \times 0 \end{bmatrix}$$

$$= \begin{bmatrix} 1 & 0 \\ 0 & 1 \end{bmatrix} = I$$

and likewise we show that $A^{-1} \cdot A = I$.

Example 2
The matrices

$$A = \begin{bmatrix} \frac{1}{2} & -1 & \frac{3}{2} \\ 1 & \frac{1}{2} & -\frac{3}{2} \\ -\frac{1}{2} & \frac{1}{2} & 1 \end{bmatrix} \quad \text{and} \quad A^{-1} = \begin{bmatrix} \frac{5}{8} & \frac{7}{8} & \frac{3}{8} \\ -\frac{1}{8} & \frac{5}{8} & \frac{9}{8} \\ \frac{3}{8} & \frac{1}{8} & \frac{5}{8} \end{bmatrix}$$

are also inverses of one another. We leave the verification as an exercise.

With an inverse we can solve a system of linear equations, as follows: Let A be an n by n matrix as in (2); let

$$X = \begin{bmatrix} x_1 \\ x_2 \\ \cdot \\ \cdot \\ \cdot \\ x_n \end{bmatrix} \quad \text{and} \quad B = \begin{bmatrix} b_1 \\ b_2 \\ \cdot \\ \cdot \\ \cdot \\ b_n \end{bmatrix}$$

and consider the system of n linear equations in n unknowns

$$A \cdot X = B \tag{3}$$

If A has an inverse A^{-1}, then

$$A^{-1} \cdot (A \cdot X) = A^{-1} \cdot B$$

But

$$A^{-1} \cdot (A \cdot X) = (A^{-1} \cdot A) \cdot X = I \cdot X = X$$

so that $X = A^{-1} \cdot B$. This relation gives a value for each of the unknowns $x_1, x_2, \ldots, x_n$, and hence we have found the solution of (3).

Example 3

Solve the system of linear equations

$$\frac{1}{2}x_1 - x_2 + \frac{3}{2}x_3 = 1$$

$$x_1 + \frac{1}{2}x_2 - \frac{3}{2}x_3 = 0$$

$$-\frac{1}{2}x_1 + \frac{1}{2}x_2 + x_3 = -1$$

SOLUTION

The coefficient matrix here is the matrix A of Example 2. If we set

$$B = \begin{bmatrix} 1 \\ 0 \\ -1 \end{bmatrix} \quad \text{and} \quad X = \begin{bmatrix} x_1 \\ x_2 \\ x_3 \end{bmatrix}$$

then the solution is

$$A^{-1} \cdot B = \begin{bmatrix} \frac{5}{8} & \frac{7}{8} & \frac{3}{8} \\ -\frac{1}{8} & \frac{5}{8} & \frac{9}{8} \\ \frac{3}{8} & \frac{1}{8} & \frac{5}{8} \end{bmatrix} \cdot \begin{bmatrix} 1 \\ 0 \\ -1 \end{bmatrix} = \begin{bmatrix} \frac{1}{4} \\ -\frac{5}{4} \\ -\frac{1}{4} \end{bmatrix} = \begin{bmatrix} x_1 \\ x_2 \\ x_3 \end{bmatrix}$$

Hence, $x_1 = \frac{1}{4}$, $x_2 = -\frac{5}{4}$, $x_3 = -\frac{1}{4}$.

Calculation of the inverse matrix

We shall now show how to find the inverse of a 3 by 3 matrix with the pivotal reduction method. The reasoning is, however, general, and it applies to any size square matrix. Let

$$A = \begin{bmatrix} a_{11} & a_{12} & a_{13} \\ a_{21} & a_{22} & a_{23} \\ a_{31} & a_{32} & a_{33} \end{bmatrix}$$

be a given matrix, and let it have an inverse

$$A^{-1} = \begin{bmatrix} x_1 & x_2 & x_3 \\ y_1 & y_2 & y_3 \\ z_1 & z_2 & z_3 \end{bmatrix}$$

Our goal is to find the *unknown* elements of A^{-1} by using the elements of A. For this we use the relation

$$\begin{bmatrix} a_{11} & a_{12} & a_{13} \\ a_{21} & a_{22} & a_{23} \\ a_{31} & a_{32} & a_{33} \end{bmatrix} \cdot \begin{bmatrix} x_1 & x_2 & x_3 \\ y_1 & y_2 & y_3 \\ z_1 & z_2 & z_3 \end{bmatrix} = \begin{bmatrix} 1 & 0 & 0 \\ 0 & 1 & 0 \\ 0 & 0 & 1 \end{bmatrix} \qquad (4)$$

Here we have a system of linear equations in matrix form and, as in Section 6.3, we obtain a solution by reducing the matrix A in the left side to I. This procedure will be illustrated with examples.

Example 4
Find the inverse of the matrix

$$A = \begin{bmatrix} 1 & 0 & -1 \\ -2 & 1 & 2 \\ -1 & 2 & -2 \end{bmatrix}$$

SOLUTION
In view of (4), the augmented matrix now is

$$\left[\begin{array}{ccc|ccc} 1 & 0 & -1 & 1 & 0 & 0 \\ -2 & 1 & 2 & 0 & 1 & 0 \\ -1 & 2 & -2 & 0 & 0 & 1 \end{array} \right]$$

We apply the pivotal reduction method to this matrix in the same way as before (see page 273).

We arrive at the matrix

$$A^{-1} = \begin{bmatrix} 2 & \frac{2}{3} & -\frac{1}{3} \\ 2 & 1 & 0 \\ 1 & \frac{2}{3} & -\frac{1}{3} \end{bmatrix}$$

The matrix A^{-1} is the inverse of A, as is verified by computing $A \cdot A^{-1}$. We leave this as an exercise.

Example 5
Find the inverse of

$$A = \begin{bmatrix} 1 & 1 & 0 \\ 0 & 1 & 1 \\ 1 & 0 & 1 \end{bmatrix}$$

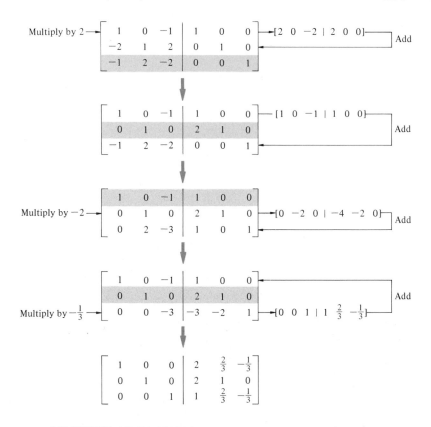

SOLUTION OF EXAMPLE 5

Beginning with the augmented matrix, we proceed as in Example 4 (see page 274).

Thus the inverse matrix is

$$A^{-1} = \begin{bmatrix} \frac{1}{2} & -\frac{1}{2} & \frac{1}{2} \\ \frac{1}{2} & \frac{1}{2} & -\frac{1}{2} \\ -\frac{1}{2} & \frac{1}{2} & \frac{1}{2} \end{bmatrix}$$

Example 6
Find the inverse of

$$A = \begin{bmatrix} -1 & 0 & 1 \\ 1 & -1 & 0 \\ 0 & 1 & -1 \end{bmatrix}$$

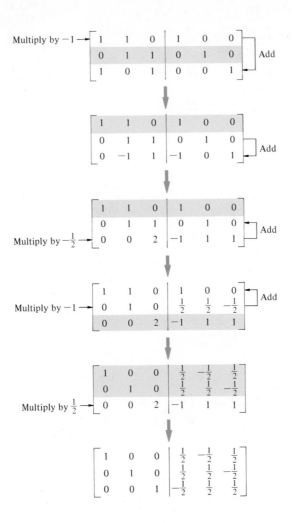

Multiply by -1 →
$$\left[\begin{array}{ccc|ccc} 1 & 1 & 0 & 1 & 0 & 0 \\ 0 & 1 & 1 & 0 & 1 & 0 \\ 1 & 0 & 1 & 0 & 0 & 1 \end{array}\right]$$ Add

↓

$$\left[\begin{array}{ccc|ccc} 1 & 1 & 0 & 1 & 0 & 0 \\ 0 & 1 & 1 & 0 & 1 & 0 \\ 0 & -1 & 1 & -1 & 0 & 1 \end{array}\right]$$ Add

↓

Multiply by $-\frac{1}{2}$ →
$$\left[\begin{array}{ccc|ccc} 1 & 1 & 0 & 1 & 0 & 0 \\ 0 & 1 & 1 & 0 & 1 & 0 \\ 0 & 0 & 2 & -1 & 1 & 1 \end{array}\right]$$ Add

↓

Multiply by -1 →
$$\left[\begin{array}{ccc|ccc} 1 & 1 & 0 & 1 & 0 & 0 \\ 0 & 1 & 0 & \frac{1}{2} & \frac{1}{2} & -\frac{1}{2} \\ 0 & 0 & 2 & -1 & 1 & 1 \end{array}\right]$$ Add

↓

Multiply by $\frac{1}{2}$ →
$$\left[\begin{array}{ccc|ccc} 1 & 0 & 0 & \frac{1}{2} & -\frac{1}{2} & \frac{1}{2} \\ 0 & 1 & 0 & \frac{1}{2} & \frac{1}{2} & -\frac{1}{2} \\ 0 & 0 & 2 & -1 & 1 & 1 \end{array}\right]$$

↓

$$\left[\begin{array}{ccc|ccc} 1 & 0 & 0 & \frac{1}{2} & -\frac{1}{2} & \frac{1}{2} \\ 0 & 1 & 0 & \frac{1}{2} & \frac{1}{2} & -\frac{1}{2} \\ 0 & 0 & 1 & -\frac{1}{2} & \frac{1}{2} & \frac{1}{2} \end{array}\right]$$

SOLUTION OF EXAMPLE 6

$$\left[\begin{array}{ccc|ccc} -1 & 0 & 1 & 1 & 0 & 0 \\ 1 & -1 & 0 & 0 & 1 & 0 \\ 0 & 1 & -1 & 0 & 0 & 1 \end{array}\right]$$ Add

↓

$$\left[\begin{array}{ccc|ccc} -1 & 0 & 1 & 1 & 0 & 0 \\ 0 & -1 & 1 & 1 & 1 & 0 \\ 0 & 1 & -1 & 0 & 0 & 1 \end{array}\right]$$ Add

↓

$$\left[\begin{array}{ccc|ccc} -1 & 0 & 1 & 1 & 0 & 0 \\ 0 & -1 & 1 & 1 & 1 & 0 \\ 0 & 0 & 0 & 1 & 1 & 1 \end{array}\right]$$

Having a row of zeros, we see that the coefficient matrix *cannot* be reduced to the identity matrix. Hence, A has *no* inverse.

EXERCISES

In Exercises 1–14 find the inverses for the given matrices when the inverses exist.

1. $\begin{bmatrix} -1 & 0 \\ 1 & -1 \end{bmatrix}$

2. $\begin{bmatrix} 0 & 1 \\ 3 & 0 \end{bmatrix}$

3. $\begin{bmatrix} \frac{1}{2} & 3 \\ -3 & -1 \end{bmatrix}$

4. $\begin{bmatrix} 2 & 3 \\ 5 & -2 \end{bmatrix}$

5. $\begin{bmatrix} 1 & 1 & 0 \\ 0 & 1 & 1 \\ 1 & 2 & 1 \end{bmatrix}$

6. $\begin{bmatrix} 2 & 1 & 3 \\ 1 & 0 & 1 \\ 4 & 1 & 2 \end{bmatrix}$

7. $\begin{bmatrix} -1 & 1 & -1 \\ -2 & 7 & 2 \\ -1 & 0 & -1 \end{bmatrix}$

8. $\begin{bmatrix} 6 & 4 & 2 \\ 3 & 1 & 5 \\ 3 & 1 & 5 \end{bmatrix}$

9. $\begin{bmatrix} 1 & -2 & 3 \\ 2 & 1 & -3 \\ -1 & 1 & 2 \end{bmatrix}$

10. $\begin{bmatrix} 1 & 2 & 1 \\ 3 & 1 & 2 \\ 1 & 1 & 1 \end{bmatrix}$

11. $\begin{bmatrix} 3 & 0 & 0 \\ 0 & 3 & 0 \\ 0 & 0 & 3 \end{bmatrix}$

12. $\begin{bmatrix} 0 & 0 & 3 \\ 0 & 3 & 0 \\ 3 & 0 & 0 \end{bmatrix}$

13. $\begin{bmatrix} 1 & 0 & 1 & 0 \\ 0 & -1 & 0 & -1 \\ -1 & 0 & 1 & 1 \\ 0 & -1 & 1 & -1 \end{bmatrix}$

14. $\begin{bmatrix} 1 & 1 & 0 & 1 \\ 0 & 1 & 1 & 0 \\ 1 & 0 & -1 & 2 \\ 0 & 1 & 0 & 1 \end{bmatrix}$

15. In Example 2, show that A^{-1} is the inverse of A.
16. In Example 4, show that A^{-1} is the inverse of A.
17. In Example 5, show that A^{-1} is the inverse of A.

6.5   SYSTEMS OF AN UNEQUAL NUMBER OF EQUATIONS AND UNKNOWNS

Up to this point we discussed systems of linear equations in which the number of equations and the number of unknowns was the same. In many applications, however, this is not the case, and in this section we look at more general situations.

Example 1

Three kinds of meat and a protein-enriched filler are to be used for sausage. The amounts of protein and fat in grams per ounce are given in Table 6.1. Is it possible to combine these ingredients into a mixture that will provide 84 grams protein and 26 grams fat per pound (16 ounces)?

TABLE 6.1. QUANTITIES ARE EXPRESSED IN GRAMS PER OUNCE

	Pork	Veal	Beef	Filler
Protein	7	8	6	4
Fat	4	2	4	0

SOLUTION

Consider a 1 pound mixture of x ounces pork, y ounces veal, z ounces beef, and t ounces filler. According to the data, we have the following system of linear equations:

$$\begin{aligned} x + y + z + t &= 16 \\ 7x + 8y + 6z + 4t &= 84 \\ 4x + 2y + 4z &= 26 \end{aligned} \qquad (5)$$

Here we have a system of three equations in four unknowns. We proceed to solve it with the pivotal reduction method, using x as a pivot (see page 277).

With the last matrix on page 277, we reduced the original system of linear equations to the equivalent system:

$$\begin{aligned} x - 6t &= -50 \\ y + 2t &= 19 \\ z + 5t &= 47 \end{aligned}$$

Expressing x, y, and z in terms of t, and using the fact that $x \geq 0$, $y \geq 0$, and $z \geq 0$ (since these quantities are weights), we have

$$\begin{aligned} x &= -50 + 6t \geq 0 \\ y &= 19 - 2t \geq 0 \\ z &= 47 - 5t \geq 0 \end{aligned} \qquad (6)$$

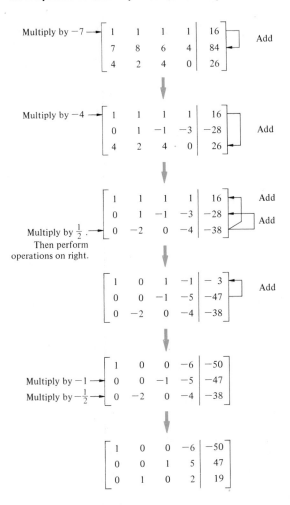

Multiply by -7 →
$$\begin{bmatrix} 1 & 1 & 1 & 1 & | & 16 \\ 7 & 8 & 6 & 4 & | & 84 \\ 4 & 2 & 4 & 0 & | & 26 \end{bmatrix}$$ Add

Multiply by -4 →
$$\begin{bmatrix} 1 & 1 & 1 & 1 & | & 16 \\ 0 & 1 & -1 & -3 & | & -28 \\ 4 & 2 & 4 & 0 & | & 26 \end{bmatrix}$$ Add

$$\begin{bmatrix} 1 & 1 & 1 & 1 & | & 16 \\ 0 & 1 & -1 & -3 & | & -28 \\ 0 & -2 & 0 & -4 & | & -38 \end{bmatrix}$$ Add Add

Multiply by $\frac{1}{2}$. Then perform operations on right.

$$\begin{bmatrix} 1 & 0 & 1 & -1 & | & 3 \\ 0 & 0 & -1 & -5 & | & -47 \\ 0 & -2 & 0 & -4 & | & -38 \end{bmatrix}$$ Add

Multiply by -1 →
Multiply by $-\frac{1}{2}$ →
$$\begin{bmatrix} 1 & 0 & 0 & -6 & | & -50 \\ 0 & 0 & -1 & -5 & | & -47 \\ 0 & -2 & 0 & -4 & | & -38 \end{bmatrix}$$

$$\begin{bmatrix} 1 & 0 & 0 & -6 & | & -50 \\ 0 & 0 & 1 & 5 & | & 47 \\ 0 & 1 & 0 & 2 & | & 19 \end{bmatrix}$$

These inequalities give

$$t \geq \frac{25}{3} = 8.333 \cdots$$

$$t \leq \frac{19}{2} = 9.5$$

$$t \leq \frac{47}{5} = 9.4$$

We see that these inequalities are satisfied by values

$$8.333 \cdots = \frac{25}{3} \leq t \leq \frac{47}{5} = 9.4$$

Hence, the system of linear equations (5) has a solution (in fact, infinitely many solutions), and so the answer to the problem is *yes*. A convenient choice for t is 9, and substituting this into (6) gives the solution

$$x = 4 \qquad y = 1 \qquad z = 2 \qquad t = 9$$

Note, however, that this solution is not unique.

There are rules for deciding when a system of m linear equations in n unknowns has a solution. It is preferable, however, especially when dealing with small systems, to work this out directly. The pivotal reduction method is suitable for this because it works equally well when $n > m$, $n = m$, or $n < m$. It will give a solution when it exists, and will show when no solution exists.

Example 2

Referring to Example 1: Suppose only veal and beef are to be used in the sausage mix. Is it possible to combine these meats into a mixture that will provide 75 grams protein and 15 grams fat per pound (16 ounces)?

SOLUTION

If x ounces of veal and y ounces of beef are used, then we have the system of linear equations

$$\begin{aligned} x + y &= 16 \\ 8x + 6y &= 75 \\ 2x + 4y &= 15 \end{aligned} \qquad (7)$$

Solving this in the usual way (see page 279) yields the following:

The last matrix implies the absurd result $0 = -70$. This tells us that the system of equations (7) is inconsistent and has no solution.

Example 3

Does the system

$$\begin{aligned} x + y &= 4 \\ 17x - 4y &= 5 \\ 4x + y &= 7 \end{aligned} \qquad (8)$$

have a solution?

Multiply by -8 ⟶ $\begin{bmatrix} 1 & 1 & 16 \\ 8 & 6 & 75 \\ 2 & 4 & 15 \end{bmatrix}$ Add

⬇

Multiply by -2 ⟶ $\begin{bmatrix} 1 & 1 & 16 \\ 0 & -2 & -53 \\ 2 & 4 & 15 \end{bmatrix}$ Add

⬇

$\begin{bmatrix} 1 & 1 & 16 \\ 0 & -2 & -53 \\ 0 & 2 & -17 \end{bmatrix}$ Add

⬇

$\begin{bmatrix} 1 & 1 & 16 \\ 0 & -2 & -53 \\ 0 & 0 & -70 \end{bmatrix}$

SOLUTION OF EXAMPLE 3

We proceed as in Example 2.

Multiply by -17 ⟶ $\begin{bmatrix} 1 & 1 & 4 \\ 17 & -4 & 5 \\ 4 & 1 & 7 \end{bmatrix}$ Add

⬇

Multiply by -4 ⟶ $\begin{bmatrix} 1 & 1 & 4 \\ 0 & -21 & -63 \\ 4 & 1 & 7 \end{bmatrix}$ Add

⬇

Multiply by $-\frac{1}{7}$ ⟶ $\begin{bmatrix} 1 & 1 & 4 \\ 0 & -21 & -63 \\ 0 & -3 & -9 \end{bmatrix}$ Add

⬇

$\begin{bmatrix} 1 & 1 & 4 \\ 0 & -21 & -63 \\ 0 & 0 & 0 \end{bmatrix}$ *

Thus, system (8) reduced to the equivalent system

$$x + y = 4$$
$$ -21y = -63$$

and this has the solution $x = 1$ and $y = 3$. These values are also a solution of (8).

We observe that the augmented matrix in $(*)$ can be reduced to the form

$$\left[\begin{array}{cc|c} 1 & 0 & 1 \\ 0 & 1 & 3 \\ \hline 0 & 0 & 0 \end{array}\right]$$

In general, if the augmented matrix of a system of m linear equations in n unknowns, with $m > n$, can be reduced to the form

$$
m-n\left\{\begin{array}{c} n\left\{\begin{array}{c} \end{array}\right. \\ \end{array}\right.
\left[\begin{array}{ccccc|c}
1 & 0 & 0 & \cdots & 0 & b_1 \\
0 & 1 & 0 & \cdots & 0 & b_2 \\
0 & 0 & 1 & \cdots & 0 & b_3 \\
\cdot & \cdot & \cdot & \cdots & \cdot & \cdot \\
\cdot & \cdot & \cdot & \cdots & \cdot & \cdot \\
\cdot & \cdot & \cdot & \cdots & \cdot & \cdot \\
0 & 0 & 0 & \cdots & 1 & b_n \\
\hline
0 & 0 & 0 & \cdots & 0 & 0 \\
\cdot & & & & \cdot & \cdot \\
\cdot & & & & \cdot & \cdot \\
\cdot & & & & \cdot & \cdot \\
0 & 0 & 0 & \cdots & 0 & 0
\end{array}\right]
$$

with not all constants b_j zero, then this matrix gives the solution of the original system. To get the matrix to this particular form may require an interchange of rows. If the column of constants in this matrix contains a nonzero constant *below* the nth row, then the original system has no solution.

EXERCISES

1. In Example 1, suppose the filler contains 2 grams fat per ounce. Is it still possible to find a mixture of ingredients with 60 grams protein and 30 grams fat per pound?

Use the pivotal reduction method in Exercises 2–14 to find all existing solutions.

2. $\begin{aligned} x - y - z &= 0 \\ x \quad\quad + 3z &= 6 \\ 2y - 3z &= 0 \\ x + 2y \quad\quad &= 6 \end{aligned}$

3. $\begin{aligned} 3x + y + z &= 2 \\ -2x + y + z &= 7 \\ x - 2y + 2z &= 5 \\ 5x + 3y + z &= 2 \end{aligned}$

4. $\begin{aligned} x + y + z &= 1 \\ x - y + z &= 0 \\ x + y - z &= -1 \\ -x + y + z &= 0 \end{aligned}$

5. $\begin{aligned} x + y + z &= 9 \\ x + 2y - z &= 0 \\ x - y + 2z &= 12 \\ 2x - y + 2z &= 15 \end{aligned}$

6. $\begin{aligned} x + y + z &= 3 \\ 5x - 4y \quad\quad &= 1 \\ 3x \quad\quad + 7z &= 10 \\ -2y + 3z &= 1 \end{aligned}$

7. $\begin{aligned} x + y + z &= 2 \\ x - y + z &= 4 \\ -x - y + z &= 2 \\ x - y - z &= 0 \end{aligned}$

8. $\begin{aligned} -4x + 2y + 2z &= -1 \\ x + y - 6z &= 2 \\ 6x - y - 18z &= 0 \\ 3x - y + 6z &= 0 \end{aligned}$

9. $\begin{aligned} x + y + z &= 9 \\ 2x - y + 2z &= 15 \\ x - y + 2z &= 12 \\ x + 2y - z &= 0 \end{aligned}$

10. $\begin{aligned} x + y + z + t &= 16 \\ 8x + 4y + 7z + 3t &= 96 \\ x + 3y + 5z + 3t &= 40 \end{aligned}$

11. $\begin{aligned} x + y + z - t &= 1 \\ x + y \quad\quad + 2t &= 0 \\ y + z - t &= -1 \end{aligned}$

12. $\begin{aligned} x + y + 3z &= 8 \\ x + 2y + z &= 6 \end{aligned}$

13. $\begin{aligned} x + y + z &= 0 \\ 2x + y + 3z &= 6 \end{aligned}$

14. $\begin{aligned} 3x + y - z &= 8 \\ 5x - 2y + 2z &= -5 \end{aligned}$

6.6 COORDINATES AND LINEAR EQUATIONS IN THREE DIMENSIONS

It is convenient at times to have the ability of spatial perception, at least as far as visualizing linear equations in three variables is concerned. For this purpose we introduce here three-dimensional coordinates. Three-dimensional graphing is difficult, and you are not expected to master it. You should, however, be able to interpret the diagrams and benefit from their use as illustrations.

In Section 5.1 we saw that each point in the xy-plane could be identified with an ordered pair (x, y) of real numbers. When we con-

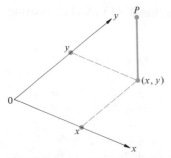

FIGURE 6.5 *Point* P *lying outside the* xy-*plane*

sider a point P not lying in this plane (see Figure 6.5) we see that an ordered pair no longer suffices to fix P, since its height, above or below the plane, has to be taken into account. Because a point in three-dimensional space is determined by a pair of xy-coordinates plus height, it is suggested that three numbers be used to identify P. This is accomplished as follows: We augment the xy-coordinate system constructed for the plane by erecting through the origin $(0, 0)$ a number line perpendicular to this plane. The number line is so erected that its origin coincides with $(0, 0)$ and it is oriented as in Figure 6.6. The new axis, customarily called the z-*axis*, is perpendicular to both the

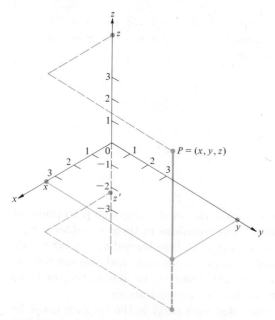

FIGURE 6.6 *Three-dimensional coordinate system*

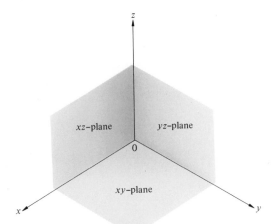

FIGURE 6.7   *The coordinate planes*

x-axis and the y-axis. In this xyz-coordinate system, each point P is identified with an ordered 3-tuple (x, y, z) of real numbers, z being positive when P lies above the xy-plane, and negative when P lies below it (see Figure 6.6). We note that if (x, y, z) and (x', y', z') represent points in 3-dimensional space, then

$$(x, y, z) = (x', y', z')$$

when and only when

$$x = x' \qquad y' = y \qquad \text{and} \qquad z = z'$$

Each pair of coordinates determines a plane, called a *coordinate plane*. The three coordinate axes give three such coordinate planes, the xy-plane, the yz-plane, and the xz-plane (see Figure 6.7).

Example 1
Let us plot the points $(1, 0, 1)$, $(0, 1, 1)$, and $(1, 1, 1)$. We observe that all three points lie one unit above the xy-plane. The point $(1, 0, 1)$ must lie in the xz-plane because the second coordinate (the y-coordinate) is zero; the point $(0, 1, 1)$ must lie in the yz-plane. As shown in Figure 6.8, the point $(1, 1, 1)$ lies one unit above each of the coordinate planes.

Linear Equations
The locus of points (x, y, z) satisfying a linear equation

$$ax + by + cz = d$$

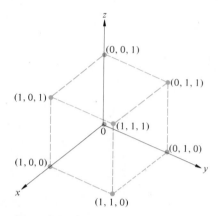

FIGURE 6.8

with a, b, and c, not all zero, is a *plane*, called the *graph* of the equation. When graphing lines in the plane we made use of the fact that a straight line is determined by any two points on it. We now make use of the fact that

a plane is uniquely determined by any three points on it.

Example 2
Find the graph of the equation $x + y + z = 1$.

SOLUTION

Three convenient points satisfying this equation are $(1, 0, 0)$, $(0, 1, 0)$, and $(0, 0, 1)$. A portion of the plane determined by these points is given in Figure 6.9.

Example 3
Find the graph of the equation $x + y = 1$.

SOLUTION

This equation can be written as

$$x + y + 0 \cdot z = 1$$

We see that this equation is satisfied by all values of z, but only by the values x and y lying on the line $x + y = 1$. The graph is therefore a plane containing this line and perpendicular to the xy-plane. Three convenient points on the plane are $(1, 0, 0)$, $(0, 1, 0)$, and $(1, 0, 1)$ (see Figure 6.10).

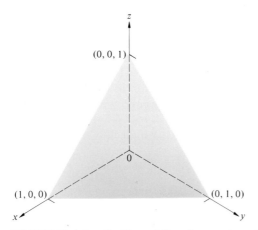

FIGURE 6.9 *Portion of the plane* x + y + z = *1*

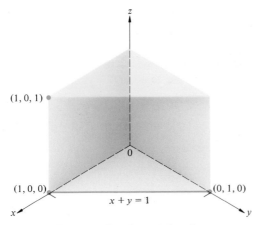

FIGURE 6.10 *Portion of the plane* x + y = *1*

Two planes can either be parallel or coincident, or else they intersect in a line. Since planes are defined by means of linear equations, we know how to determine if they intersect, and how to find the line of intersection when they do. Below we show how to do this geometrically.

Example 4
Determine the line of intersection of the planes

$$\frac{1}{2}x + \frac{1}{3}y + z = 1$$

$$x + y = 1$$

SOLUTION

Since a line is determined by any two points on it, we can find the line of intersection by choosing two convenient points (x_0, y_0, z_0) satisfying the given system of linear equations. Taking $x = 1$ in the second equation gives $y = 0$, and a substitution in the first equation fives $z = \frac{1}{2}$. Hence, $(1, 0, \frac{1}{2})$ is one of our points. Similarly, $y = 1$ in the second equation gives $x = 0$, and our second point is found from the first equation to be $(0, 1, \frac{2}{3})$. The planes and line of intersection are plotted in Figure 6.11.

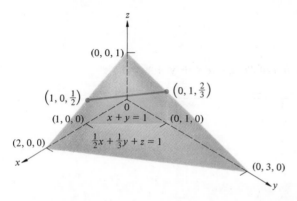

FIGURE 6.11   *The intersection of the planes* $\frac{1}{2}$x $+ \frac{1}{3}$y $+$ z $=$ *1 and* x $+$ y $=$ *1*

Example 5
How do the planes

$$\frac{1}{3}x + y + z = 1 \quad \text{and} \quad x + 2y + \frac{1}{2}z = 1$$

intersect?

SOLUTION

The first plane is determined by the three points $(3, 0, 0)$, $(0, 1, 0)$, and $(0, 0, 1)$; the second plane is determined by the three points $(1, 0, 0)$, $(0, \frac{1}{2}, 0)$, and $(0, 0, 2)$. (See Figure 6.12.) Taking $y = 0$ gives the two equations

$$\frac{1}{3}x + z = 1 \quad \text{and} \quad x + \frac{1}{2}z = 1$$

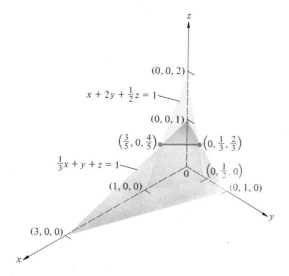

FIGURE 6.12   *The intersection of the planes $\frac{1}{3}x + y + z = 1$ and $x + 2y + \frac{1}{2}z = 1$*

Using the method in Section 5.3 or Section 6.3, we find the solution to be $x = \frac{3}{5}$ and $z = \frac{4}{5}$. Hence, the point $(\frac{3}{5},\ 0,\ \frac{4}{5})$ lies in both planes. Taking $x = 0$ gives the equations

$$y + z = 1 \quad \text{and} \quad 2y + \frac{1}{2}z = 1$$

This system has the solution $y = \frac{1}{3}$ and $z = \frac{2}{3}$, and hence also the point $(0,\ \frac{1}{3},\ \frac{2}{3})$ lies in both planes (see Figure 6.12).

EXERCISES

1. Plot and label the following points in Figure 6.13.

   | (2, 0, 0) | (0, 4, 0) | (0, 0, 3) | (2, 4, 0) |
   | (2, 0, 3) | (0, 4, 3) | (2, 4, 3) | |

2. The upper face of the brick in Figure 6.14 lies in the xy-plane as shown. Find the coordinates of all vertices.
3. What is the locus of all points $(x, 3, z)$?
4. What is the locus of all points $(x, 3, 3)$?

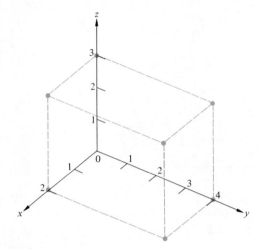

FIGURE 6.13

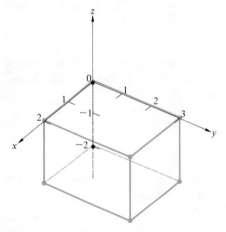

FIGURE 6.14

7
MARKOV CHAINS

7

The type of problem dealt with in this chapter concerns systems which change with time and take on different states. The system may be the economy, a line at a ticket counter, a population, and so on. A finite Markov chain is a probabilistic process which determines the changes of a system from one state to another state out of a finite number of possible choices. An important feature of the process is that a change, called *transition,* from a given state to another, depends on the given state but on no earlier ones. More precisely, we introduce the following definition.

DEFINITION 1

An experiment with a finite sequence of trials is a (finite) *Markov chain* if

    (1)   Each trial has a finite number of possible outcomes.
    (2)   The outcome of a given trial depends on the outcome of the immediately preceding trial, but on no earlier trials.

We shall explain this further while introducing the basic ideas and terminology. This is done in the following examples.

Random-Walk Problem

A man is standing in one of six positions on a straight line (see Figure 7.1). He can step from one position to another in unit steps, that is, without intermediate positions. The man steps to the right with probability p, and to the left with probability $q = 1 - p$. The positions, which change with time, are called *states*; the first and last state (a_1 and a_6 in Figure 7.1) are called *border states*, and once reached, the man remains there indefinitely. The process is said to have an *absorb-*

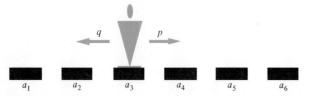

FIGURE 7.1

ing barrier at a_1 and a_6. The type of question we are interested in is this: If the process starts at state a_3, for example, what is the probability that it will be at state a_5 after four steps?

The problem is an example of a *Markov chain* or *Markov process*. It is presented schematically as in Figure 7.2, where the process starts from state a_3. You may recall similar diagrams from Example 2 and Exercise 7 in Section 4.3, but there is a basic difference here: The scheme in Figure 7.2 repeats itself indefinitely. A "closed form" of the process is given in Figure 7.3, which is called a *transition diagram*. The arrows indicate the possible transitions from one state to another. The diagram helps to explain the use of the term *chain*. In both Figures

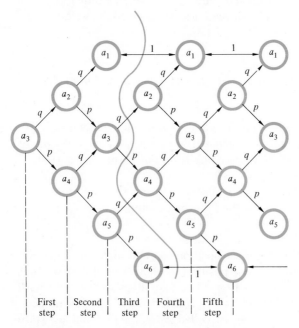

FIGURE 7.2 *Schematic diagram for the random-walk problem with absorbing barrier at* a_1 *and* a_6

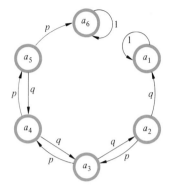

FIGURE 7.3   *Transition diagram for the random-walk problem with absorbing barrier at a_1 and a_6*

7.2 and 7.3 we noted the *transition probabilities* from one state to another.

To associate the transition probabilities with the states in a form which is useful for manipulations, we make use of matrix notation. In the (i, j)-position of a 6 by 6 matrix, we put the transition probability from state a_i to state a_j. The resulting matrix T is given below (1).

To state

$$
\begin{array}{c}
\text{From} \\
\text{state}
\end{array}
\begin{array}{c}
\\
a_1 \\
a_2 \\
a_3 \\
a_4 \\
a_5 \\
a_6
\end{array}
\begin{array}{c}
\begin{array}{cccccc}
a_1 & a_2 & a_3 & a_4 & a_5 & a_6
\end{array} \\
\left[
\begin{array}{cccccc}
1 & 0 & 0 & 0 & 0 & 0 \\
q & 0 & p & 0 & 0 & 0 \\
0 & q & 0 & p & 0 & 0 \\
0 & 0 & q & 0 & p & 0 \\
0 & 0 & 0 & q & 0 & p \\
0 & 0 & 0 & 0 & 0 & 1
\end{array}
\right]
\end{array} = T \tag{1}
$$

A zero entry in the matrix indicates that the transition between two states cannot occur. Notice in particular that the first and last rows correspond to the boundary states. The entry 1 tells us that these are terminal states, since 1 is the probability that the process will remain in that state. The second row gives the transition probability q from a_2 to a_1, and the transition probability p from a_2 to a_3, and so on. Finally, we note that

Property (1)   The entries of the matrix T are nonnegative.
Property (2)   The sum of each row of T equals 1.

This matrix is called a *transition matrix* or *stochastic matrix,* and the two properties noted are true of all transition matrices. Property (1) is necessary because the entries are probabilities; property (2) is true because a row gives the probabilities of all possible transitions (outcomes) of a given state.

Example 1

Suppose the man decides each move by tossing a coin. The probability of moving right or left is then $p = q = \frac{1}{2}$ and the transition matrix becomes

To state

$$
\begin{array}{c c}
 & \begin{array}{c c c c c c} a_1 & a_2 & a_3 & a_4 & a_5 & a_6 \end{array} \\
\begin{array}{c} \\ \\ \text{From} \\ \text{state} \\ \\ \end{array}
\begin{array}{c} a_1 \\ a_2 \\ a_3 \\ a_4 \\ a_5 \\ a_6 \end{array}
&
\left[
\begin{array}{c c c c c c}
1 & 0 & 0 & 0 & 0 & 0 \\
\frac{1}{2} & 0 & \frac{1}{2} & 0 & 0 & 0 \\
0 & \frac{1}{2} & 0 & \frac{1}{2} & 0 & 0 \\
0 & 0 & \frac{1}{2} & 0 & \frac{1}{2} & 0 \\
0 & 0 & 0 & \frac{1}{2} & 0 & \frac{1}{2} \\
0 & 0 & 0 & 0 & 0 & 1
\end{array}
\right] = T
\end{array}
$$

Example 2

Starting from state a_3, what is the probability that the man will reach state a_5 after four transitions if $p = q = \frac{1}{2}$?

SOLUTION

This problem can be solved with a direct count of the possible transitions. From Figure 7.3 (or Figure 7.2) we find the following possible chains:

$$C_1: \quad a_3 \xrightarrow{q} a_2 \xrightarrow{p} a_3 \xrightarrow{p} a_4 \xrightarrow{p} a_5$$

$$C_2: \quad a_3 \xrightarrow{p} a_4 \xrightarrow{q} a_3 \xrightarrow{p} a_4 \xrightarrow{p} a_5$$

$$C_3: \quad a_3 \xrightarrow{p} a_4 \xrightarrow{p} a_5 \xrightarrow{q} a_4 \xrightarrow{p} a_5$$

This is a stochastic process and the probabilities of these possible outcomes are

$$P(C_1) = q \times p \times p \times p = p^3 q$$
$$P(C_2) = p \times q \times p \times p = p^3 q$$
$$P(C_3) = p \times p \times q \times p = p^3 q$$

The probability of reaching state a_5 after four transitions is therefore

$$P(C_1) + P(C_2) + P(C_3) = 3p^3q$$

Using $p = q = \frac{1}{2}$ gives for the answer $\frac{3}{16}$.

Suppose we modify the random-walk problem so that from state a_1 the man moves to state a_2, and from state a_6 he moves to state a_5. In this case the process is said to have a *reflecting barrier* at a_1 and a_6. The transition matrix will show this modification in the first and last rows, since the probability 1 is now assigned to the transition from state a_1 to a_2, and from a_6 to a_5. The new transition matrix is given below (2).

To state

$$
\begin{array}{c}
\\
\\
\text{From} \\
\text{state}
\end{array}
\begin{array}{c}
\\
a_1 \\
a_2 \\
a_3 \\
a_4 \\
a_5 \\
a_6
\end{array}
\begin{array}{cccccc}
a_1 & a_2 & a_3 & a_4 & a_5 & a_6 \\
\left[\begin{array}{cccccc}
0 & 1 & 0 & 0 & 0 & 0 \\
q & 0 & p & 0 & 0 & 0 \\
0 & q & 0 & p & 0 & 0 \\
0 & 0 & q & 0 & p & 0 \\
0 & 0 & 0 & q & 0 & p \\
0 & 0 & 0 & 0 & 1 & 0
\end{array}\right] = Q
\end{array}
\qquad (2)
$$

The transition diagram of this matrix is given in Figure 7.4. Also this matrix is, of course, stochastic, and to fix ideas, we state the following definition.

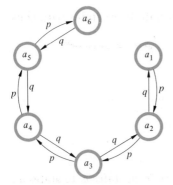

FIGURE 7.4   *Transition diagram for the random-walk problem with reflecting barrier at a_1 and a_6*

DEFINITION 2

The n by n matrix $A = [a_{ij}]$ is *stochastic* if

(1)　$a_{ij} \geq 0$ for $i = 1, 2, \ldots, n$ and $j = 1, 2, \ldots, n$
(2)　$a_{i1} + a_{i2} + \cdots + a_{in} = 1$ for $i = 1, 2, \ldots, n$

That is, a stochastic matrix consists of nonnegative elements such that the sum of the entries of each row is 1.

Example 3

Consider the matrices

$$Q = \begin{bmatrix} \frac{1}{6} & \frac{2}{6} & \frac{3}{6} \\ \frac{1}{2} & 0 & \frac{1}{2} \\ \frac{1}{4} & \frac{1}{2} & \frac{1}{4} \end{bmatrix} \qquad R = \begin{bmatrix} \frac{1}{3} & -\frac{1}{3} & \frac{1}{3} \\ 0 & \frac{1}{2} & \frac{1}{2} \\ 1 & 0 & 0 \end{bmatrix} \qquad T = \begin{bmatrix} \frac{1}{3} & \frac{1}{3} & \frac{1}{3} \\ \frac{2}{5} & \frac{1}{5} & \frac{2}{5} \\ 0 & 0 & \frac{1}{2} \end{bmatrix}$$

The matrix Q is stochastic because its elements are nonnegative, and

$$\frac{1}{6} + \frac{2}{6} + \frac{3}{6} = 1, \quad \frac{1}{2} + 0 + \frac{1}{2} = 1, \quad \text{and} \quad \frac{1}{4} + \frac{1}{2} + \frac{1}{4} = 1$$

The matrix R is not stochastic because $-\frac{1}{3} < 0$.
The matrix T is not stochastic because $0 + 0 + \frac{1}{2} \neq 1$.

Example 4

Draw the transition diagram for the matrix Q in Example 2.

SOLUTION

Let the states be a_1, a_2, a_3. We associate these states with Q as follows:

To state

		a_1	a_2	a_3
From	a_1	$\frac{1}{6}$	$\frac{2}{6}$	$\frac{3}{6}$
state	a_2	$\frac{1}{2}$	0	$\frac{1}{2}$
	a_3	$\frac{1}{4}$	$\frac{1}{2}$	$\frac{1}{4}$

We can read off the transition probabilities from states a_i to states a_j. We see in particular that every state can be reached from every other state, but a_2 cannot be followed by a_2. The diagram is given in Figure 7.5.

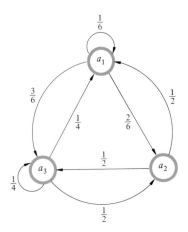

FIGURE 7.5

Gambler's-Ruin Problem

Suppose you and I play a coin-toss game. Each of us has $2.00. On Heads you win $1.00, and on Tails I win $1.00, and the game ends when either one of us has the $4.00.

Suppose further that your probability for winning a turn is p, and thus $q = 1 - p$ is the probability for losing a turn. For an unbiased coin, $p = \frac{1}{2}$ and the game is *fair*. When the coin is biased and $p > \frac{1}{2}$, the game is *favorable* to you, and when $p < \frac{1}{2}$ the game is *unfavorable* to you.

Let the states of your balance be designated

$$a_1 = 0 \qquad a_2 = 1 \qquad a_3 = 2 \qquad a_4 = 3 \qquad a_5 = 4$$

The transition diagram for this problem is given in Figure 7.6. Observe that your balance at any turn is determined by a random walk among the states $a_1, a_2, \ldots, a_5$. The transition matrix G is given below (3).

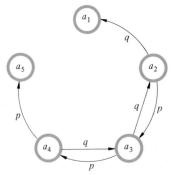

FIGURE 7.6   *Transition diagram for the matrix G in (3)*

To state

$$
\begin{array}{c}
 & \begin{array}{ccccc} a_1 & a_2 & a_3 & a_4 & a_5 \end{array} \\
\begin{matrix} a_1 \\ a_2 \\ a_3 \\ a_4 \\ a_5 \end{matrix}
\begin{bmatrix}
1 & 0 & 0 & 0 & 0 \\
q & 0 & p & 0 & 0 \\
0 & q & 0 & p & 0 \\
0 & 0 & q & 0 & p \\
0 & 0 & 0 & 0 & 1
\end{bmatrix} = G
\end{array}
\qquad (3)
$$

From state

Note that, except for the dimension, this matrix is similar to the matrix in (1). Note also that the *border states* a_1 and a_5 are *absorbing barriers*.

Example 5
Starting with \$2.00, what is the probability that you will have \$3.00 after three tosses of an unbiased coin?

SOLUTION

The starting state in this problem is a_3, and we look for all chains of three successive transitions which terminate in the state a_4. From Figure 7.6 we find the following possibilities:

$$
G_1: \quad a_3 \xrightarrow{q} a_2 \xrightarrow{p} a_3 \xrightarrow{p} a_4
$$

$$
G_2: \quad a_3 \xrightarrow{p} a_4 \xrightarrow{q} a_3 \xrightarrow{p} a_4
$$

Again, this is a stochastic process and the probabilities of these possible transitions are

$$
P(G_1) = q \times p \times p = p^2 q
$$
$$
P(G_2) = p \times q \times p = p^2 q
$$

The probability of being in state a_4 after three transitions is therefore

$$
P(G_1) + P(G_2) = 2p^2 q
$$

Using the value $p = q = \frac{1}{2}$ gives $\frac{1}{4}$ for the answer.

Example 6
Starting with \$2.00, what is the probability that you will lose your \$2.00 after four tosses of an unbiased coin?

SOLUTION

As in Example 5, we start from state a_3, but now we want to be in state a_1 after four transitions. Inspecting Figure 7.6, we see that there are only two such possible chains of transition, namely,

$$a_3 \xrightarrow{q} a_2 \xrightarrow{p} a_3 \xrightarrow{q} a_2 \xrightarrow{q} a_1$$

$$a_3 \xrightarrow{p} a_4 \xrightarrow{q} a_3 \xrightarrow{q} a_2 \xrightarrow{q} a_1$$

These transition chains have probabilities

$$q \times p \times q \times q = pq^3$$
$$p \times q \times q \times q = pq^3$$

The probability of the occurrence is, therefore, $pq^3 + pq^3 = 2pq^3$, and $p = q = \frac{1}{2}$ gives $\frac{1}{8}$ for the answer.

Thus far we used matrix notation for "bookkeeping" purposes only. In the coming sections we shall learn how to use matrices for determining the state of a system after a given number of transitions.

EXERCISES

1. This exercise refers to the random-walk problem with absorbing barrier at a_1 and a_6 (see Figure 7.3).

    (a) Starting from state a_2, what is the probability that the man will be in state a_1 after one transition? After three transitions?

    (b) Starting from state a_2, what is the probability that the man will be in state a_2 after four transitions if $p = \frac{1}{4}$ and $q = \frac{3}{4}$?

2. This exercise refers to the random-walk problem with reflecting barrier at a_1 and a_6 (see Figure 7.4).

    (a) Starting from state a_3, what is the probability of being in state a_1 after four transitions?

    (b) Starting from state a_1, what is the probability of being in state a_1 after six transitions if $p = q = \frac{1}{2}$?

3. This exercise refers to the gambler's-ruin problem (see Figure 7.6).

(a) What is the probability that you will reach state a_4 from state a_2 after seven transitions?

(b) Starting with \$2.00, what is the probability that you will have \$2.00 after four tosses of an unbiased coin?

In Exercises 4–13, draw transition diagrams for the given matrices.

4. $\begin{bmatrix} 0 & 1 \\ 1 & 0 \end{bmatrix}$

5. $\begin{bmatrix} \frac{1}{2} & \frac{1}{2} \\ \frac{1}{2} & \frac{1}{2} \end{bmatrix}$

6. $\begin{bmatrix} \frac{1}{4} & \frac{3}{4} \\ 1 & 0 \end{bmatrix}$

7. $\begin{bmatrix} 0 & 0 & 1 \\ \frac{1}{2} & 0 & \frac{1}{2} \\ 1 & 0 & 0 \end{bmatrix}$

8. $\begin{bmatrix} \frac{1}{3} & \frac{1}{3} & \frac{1}{3} \\ \frac{1}{2} & \frac{1}{4} & \frac{1}{4} \\ 0 & \frac{1}{2} & \frac{1}{2} \end{bmatrix}$

9. $\begin{bmatrix} 0 & \frac{1}{2} & \frac{1}{2} \\ \frac{1}{2} & 0 & \frac{1}{2} \\ \frac{1}{2} & \frac{1}{2} & 0 \end{bmatrix}$

10. $\begin{bmatrix} 0 & \frac{1}{2} & \frac{1}{2} \\ 0 & 0 & 1 \\ 0 & 1 & 0 \end{bmatrix}$

11. $\begin{bmatrix} 0 & \frac{1}{3} & \frac{1}{3} & \frac{1}{3} \\ \frac{1}{3} & 0 & \frac{1}{3} & \frac{1}{3} \\ \frac{1}{3} & \frac{1}{3} & 0 & \frac{1}{3} \\ \frac{1}{3} & \frac{1}{3} & \frac{1}{3} & 0 \end{bmatrix}$

12. $\begin{bmatrix} 0 & \frac{1}{2} & 0 & \frac{1}{2} \\ \frac{1}{2} & 0 & \frac{1}{2} & 0 \\ 0 & \frac{1}{2} & 0 & \frac{1}{2} \\ \frac{1}{2} & 0 & \frac{1}{2} & 0 \end{bmatrix}$

13. $\begin{bmatrix} 0 & \frac{1}{4} & \frac{1}{2} & \frac{1}{4} \\ 0 & 0 & 1 & 0 \\ 1 & 0 & 0 & 0 \\ 0 & 0 & 1 & 0 \end{bmatrix}$

In Exercises 14–18, give the transition matrix for each transition diagram.

14.

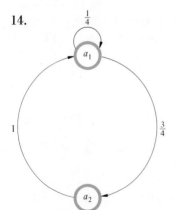

15.

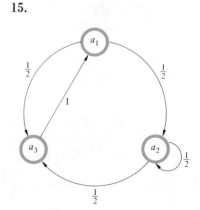

16.   17.

18.

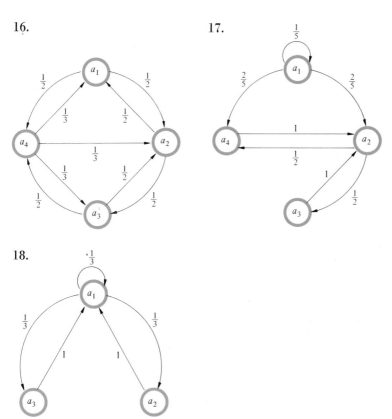

7.2   PROBABILITY VECTORS AND TRANSITION MATRICES

Consider a system with three states a_1, a_2, and a_3. Let the transition probability from state a_1 to a_1 be p_{11}, that from state a_1 to a_2 be p_{12}, and in general let the transition probability from state a_i to a_j be p_{ij}. Then the transition matrix of the system is

$$\text{To state}$$

$$\begin{array}{c} \\ \text{From} \\ \text{state} \end{array} \begin{array}{c} \\ a_1 \\ a_2 \\ a_3 \end{array} \overset{\begin{array}{ccc} a_1 & a_2 & a_3 \end{array}}{\begin{bmatrix} p_{11} & p_{12} & p_{13} \\ p_{21} & p_{22} & p_{23} \\ p_{31} & p_{32} & p_{33} \end{bmatrix}} = T \tag{4}$$

In this matrix, $p_{ij} \geq 0$ for $i = 1, 2, 3$ and $j = 1, 2, 3$, and we recall that $p_{i1} + p_{i2} + p_{i3} = 1$ for $i = 1, 2, 3$. Each row of the matrix T is a *probability vector*. A general definition follows.

DEFINITION 1

A row vector $[u_1 \quad u_2 \quad \cdots \quad u_n]$ is a *probability vector* if

(1) $u_i \geq 0$ for $i = 1, 2, \ldots, n$
(2) $u_1 + u_2 + \cdots + u_n = 1$

Example 1

The following are examples of probability vectors:

$$A = [1 \quad 0 \quad 0 \quad 0]$$
$$B = [\tfrac{1}{2} \quad 0 \quad \tfrac{1}{2} \quad 0]$$
$$C = [\tfrac{1}{4} \quad \tfrac{1}{4} \quad \tfrac{1}{4} \quad \tfrac{1}{4}]$$

Now, we know that the size of a transition matrix does not change as its system undergoes transitions from one state to another. We want to find a formula for determining the transition matrices after successive steps (transitions). First we must decide, however, how to start the Markov process. We suppose that the states a_1, a_2, and a_3 have initial probabilities $p_1^{(0)}$, $p_2^{(0)}$, and $p_3^{(0)}$, respectively. These initial probabilities are written as a probability vector

$$p^{(0)} = [p_1^{(0)} \quad p_2^{(0)} \quad p_3^{(0)}]$$

The *k-step probability vector* resulting from k transitions will be designated

$$p^{(k)} = [p_1^{(k)} \quad p_2^{(k)} \quad p_3^{(k)}] \tag{5}$$

From the first column of the matrix T in (4) we see that the state a_1 can be reached from any one of the states a_1, a_2, a_3 with respective transition probabilities p_{11}, p_{21}, p_{31}. The probability of being in state a_1 after one step is therefore

$$p_1^{(1)} = p_1^{(0)}p_{11} + p_2^{(0)}p_{21} + p_3^{(0)}p_{31}$$

From the second column of T we can compute $p_2^{(1)}$, and from the third column we can compute $p_3^{(1)}$. This results in the system of equations

$$p_1^{(1)} = p_1^{(0)}p_{11} + p_2^{(0)}p_{21} + p_3^{(0)}p_{31}$$
(probability of being in state a_1)

$$p_2^{(1)} = p_1^{(0)}p_{12} + p_2^{(0)}p_{22} + p_3^{(0)}p_{32}$$
(probability of being in state a_2)

$$p_3^{(1)} = p_1^{(0)}p_{13} + p_2^{(0)}p_{23} + p_3^{(0)}p_{33}$$
(probability of being in state a_3)

Examining this system of equations and recalling from Section 6.2 the definition of matrix multiplication, we realize that this system can be written as the matrix equation

$$[p_1{}^{(1)} \quad p_2{}^{(1)} \quad p_3{}^{(1)}] = [p_1{}^{(0)} \quad p_2{}^{(0)} \quad p_3{}^{(0)}] \cdot \begin{bmatrix} p_{11} & p_{12} & p_{13} \\ p_{21} & p_{22} & p_{23} \\ p_{31} & p_{32} & p_{33} \end{bmatrix}$$

Starting now with the vector $[p_1{}^{(k-1)} \quad p_2{}^{(k-1)} \quad p_3{}^{(k-1)}]$, we can use the above argument to obtain the formula

$$[p_1{}^{(k)} \quad p_2{}^{(k)} \quad p_3{}^{(k)}] = [p_1{}^{(k-1)} \quad p_2{}^{(k-1)} \quad p_3{}^{(k-1)}] \cdot \begin{bmatrix} p_{11} & p_{12} & p_{13} \\ p_{21} & p_{22} & p_{23} \\ p_{31} & p_{32} & p_{33} \end{bmatrix}$$

In abbreviated form, this equation can be written as

$$p^{(k)} = p^{(k-1)}T \quad \text{for} \quad k = 1, 2, 3, \ldots \tag{6}$$

This is an *iteration formula,* and from it we obtain the relations

$$p^{(1)} = p^{(0)}T$$
$$p^{(2)} = p^{(1)}T = (p^{(0)}T)T = p^{(0)}T^2$$
$$p^{(3)} = p^{(2)}T = (p^{(0)}T^2)T = p^{(0)}T^3$$

and in general,

$$p^{(k)} = p^{(0)}T^k \quad \text{for} \quad k = 1, 2, 3, \ldots \tag{7}$$

The matrix T^k is, of course, the k-fold product

$$T^k = \underbrace{T \cdot T \cdot \cdots \cdot T}_{k \text{ times}}$$

Example 2
Consider the transition matrix

$$T = \begin{bmatrix} 1 & 0 \\ \frac{1}{2} & \frac{1}{2} \end{bmatrix}$$

Find the probability vectors $p^{(k)} = [p_1^{(k)} \quad p_2^{(k)}]$ for $k = 1, 2, 3, 4$ when $p^{(0)} = [0 \quad 1]$.

SOLUTION 1

We first solve this problem using Formula (6). A direct calculation gives

$$p^{(1)} = [0 \quad 1] \cdot \begin{bmatrix} 1 & 0 \\ \frac{1}{2} & \frac{1}{2} \end{bmatrix} = [\frac{1}{2} \quad \frac{1}{2}]$$

$$p^{(2)} = [\frac{1}{2} \quad \frac{1}{2}] \cdot \begin{bmatrix} 1 & 0 \\ \frac{1}{2} & \frac{1}{2} \end{bmatrix} = [\frac{3}{4} \quad \frac{1}{4}]$$

$$p^{(3)} = [\frac{3}{4} \quad \frac{1}{4}] \cdot \begin{bmatrix} 1 & 0 \\ \frac{1}{2} & \frac{1}{2} \end{bmatrix} = [\frac{7}{8} \quad \frac{1}{8}]$$

$$p^{(4)} = [\frac{7}{8} \quad \frac{1}{8}] \cdot \begin{bmatrix} 1 & 0 \\ \frac{1}{2} & \frac{1}{2} \end{bmatrix} = [\frac{15}{16} \quad \frac{1}{16}]$$

SOLUTION 2

We now solve the problem using Formula (7). For this we calculate the powers T^2, T^3, and T^4.

$$T^2 = \begin{bmatrix} 1 & 0 \\ \frac{1}{2} & \frac{1}{2} \end{bmatrix} \cdot \begin{bmatrix} 1 & 0 \\ \frac{1}{2} & \frac{1}{2} \end{bmatrix} = \begin{bmatrix} 1 \times 1 + 0 \times \frac{1}{2} & 1 \times 0 + 0 \times \frac{1}{2} \\ \frac{1}{2} \times 1 + \frac{1}{2} \times \frac{1}{2} & \frac{1}{2} \times 0 + \frac{1}{2} \times \frac{1}{2} \end{bmatrix}$$

$$= \begin{bmatrix} 1 & 0 \\ \frac{3}{4} & \frac{1}{4} \end{bmatrix}$$

$$T^3 = \begin{bmatrix} 1 & 0 \\ \frac{3}{4} & \frac{1}{4} \end{bmatrix} \cdot \begin{bmatrix} 1 & 0 \\ \frac{1}{2} & \frac{1}{2} \end{bmatrix} = \begin{bmatrix} 1 & 0 \\ \frac{7}{8} & \frac{1}{8} \end{bmatrix}$$

$$T^4 = \begin{bmatrix} 1 & 0 \\ \frac{7}{8} & \frac{1}{8} \end{bmatrix} \cdot \begin{bmatrix} 1 & 0 \\ \frac{1}{2} & \frac{1}{2} \end{bmatrix} = \begin{bmatrix} 1 & 0 \\ \frac{15}{16} & \frac{1}{16} \end{bmatrix}$$

Hence,

$$p^{(1)} = [0 \quad 1] \cdot \begin{bmatrix} 1 & 0 \\ \frac{1}{2} & \frac{1}{2} \end{bmatrix} = [\frac{1}{2} \quad \frac{1}{2}]$$

$$p^{(2)} = [0 \quad 1] \cdot \begin{bmatrix} 1 & 0 \\ \frac{3}{4} & \frac{1}{4} \end{bmatrix} = [\frac{3}{4} \quad \frac{1}{4}]$$

$$p^{(3)} = [0 \quad 1] \cdot \begin{bmatrix} 1 & 0 \\ \frac{7}{8} & \frac{1}{8} \end{bmatrix} = [\frac{7}{8} \quad \frac{1}{8}]$$

$$p^{(4)} = [0 \quad 1] \cdot \begin{bmatrix} 1 & 0 \\ \frac{15}{16} & \frac{1}{16} \end{bmatrix} = [\frac{15}{16} \quad \frac{1}{16}]$$

We observe the following: Let the transition probability from state a_i to state a_j after k steps be $p_{ij}{}^{(k)}$ for $k = 1, 2, 3, \ldots$; let the corresponding k-step transition matrix be

$$T^{(k)} = \begin{bmatrix} p_{11}{}^{(k)} & p_{12}{}^{(k)} & p_{13}{}^{(k)} \\ p_{21}{}^{(k)} & p_{22}{}^{(k)} & p_{23}{}^{(k)} \\ p_{31}{}^{(k)} & p_{32}{}^{(k)} & p_{33}{}^{(k)} \end{bmatrix}$$

Then, by the above,

$$T^{(k)} = T^k \quad \text{for} \quad k = 1, 2, 3, \ldots \tag{8}$$

That is,

---

the k-step transition matrix equals the kth power of the original transition matrix.

The preceding discussion can be applied to systems with n states $a_1, a_2, \ldots, a_n$, in which case we have for T an n by n transition matrix $T = [p_{ij}]$. Corresponding to (5), we then consider the k-step probability vector

$$p^{(k)} = [p_1{}^{(k)} \quad p_2{}^{(k)} \quad \cdots \quad p_n{}^{(k)}] \tag{5'}$$

Following the reasoning which led to (6), we obtain the same formula for the probability vector $p^{(k)}$ in (5'). The formula (6) is now the abbreviation for

$$[p_1{}^{(k)} \quad p_2{}^{(k)} \quad \cdots \quad p_n{}^{(k)}]$$
$$= [p_1{}^{(k-1)} \quad p_2{}^{(k-1)} \quad \cdots \quad p_n{}^{(k-1)}] \cdot \begin{bmatrix} p_{11} & p_{12} & \cdots & p_{1n} \\ p_{21} & p_{22} & \cdots & p_{2n} \\ \cdot & \cdot & & \cdot \\ \cdot & \cdot & & \cdot \\ \cdot & \cdot & & \cdot \\ p_{n1} & p_{n2} & \cdots & p_{nn} \end{bmatrix}$$

Example 3

Consider the matrix G from the gambler's-ruin problem of Section 7.1 for $p = q = \frac{1}{2}$:

$$G = \begin{bmatrix} 1 & 0 & 0 & 0 & 0 \\ \frac{1}{2} & 0 & \frac{1}{2} & 0 & 0 \\ 0 & \frac{1}{2} & 0 & \frac{1}{2} & 0 \\ 0 & 0 & \frac{1}{2} & 0 & \frac{1}{2} \\ 0 & 0 & 0 & 0 & 1 \end{bmatrix}$$

Find the probability vectors

$$p^{(k)} = [p_1^{(k)} \quad p_2^{(k)} \quad p_3^{(k)} \quad p_4^{(k)} \quad p_5^{(k)}]$$

for $k = 1, 2, 3, 4$, when $p^{(0)} = [0 \quad 0 \quad 1 \quad 0 \quad 0]$.

SOLUTION

Here we use the iteration Formula (6), since it is easier than calculating the powers G^2, G^3, and G^4, which are necessary for Formula (7).

$$p^{(1)} = [0 \quad 0 \quad 1 \quad 0 \quad 0] \cdot \begin{bmatrix} 1 & 0 & 0 & 0 & 0 \\ \frac{1}{2} & 0 & \frac{1}{2} & 0 & 0 \\ 0 & \frac{1}{2} & 0 & \frac{1}{2} & 0 \\ 0 & 0 & \frac{1}{2} & 0 & \frac{1}{2} \\ 0 & 0 & 0 & 0 & 1 \end{bmatrix} = [0 \quad \frac{1}{2} \quad 0 \quad \frac{1}{2} \quad 0]$$

$$p^{(2)} = [0 \quad \frac{1}{2} \quad 0 \quad \frac{1}{2} \quad 0] \cdot \begin{bmatrix} 1 & 0 & 0 & 0 & 0 \\ \frac{1}{2} & 0 & \frac{1}{2} & 0 & 0 \\ 0 & \frac{1}{2} & 0 & \frac{1}{2} & 0 \\ 0 & 0 & \frac{1}{2} & 0 & \frac{1}{2} \\ 0 & 0 & 0 & 0 & 1 \end{bmatrix} = [\frac{1}{4} \quad 0 \quad \frac{1}{2} \quad 0 \quad \frac{1}{4}]$$

$$p^{(3)} = [\frac{1}{4} \quad 0 \quad \frac{1}{2} \quad 0 \quad \frac{1}{4}] \cdot \begin{bmatrix} 1 & 0 & 0 & 0 & 0 \\ \frac{1}{2} & 0 & \frac{1}{2} & 0 & 0 \\ 0 & \frac{1}{2} & 0 & \frac{1}{2} & 0 \\ 0 & 0 & \frac{1}{2} & 0 & \frac{1}{2} \\ 0 & 0 & 0 & 0 & 1 \end{bmatrix} = [\frac{1}{4} \quad \frac{1}{4} \quad 0 \quad \frac{1}{4} \quad \frac{1}{4}]$$

$$p^{(4)} = [\frac{1}{4} \quad \frac{1}{4} \quad 0 \quad \frac{1}{4} \quad \frac{1}{4}] \cdot \begin{bmatrix} 1 & 0 & 0 & 0 & 0 \\ \frac{1}{2} & 0 & \frac{1}{2} & 0 & 0 \\ 0 & \frac{1}{2} & 0 & \frac{1}{2} & 0 \\ 0 & 0 & \frac{1}{2} & 0 & \frac{1}{2} \\ 0 & 0 & 0 & 0 & 1 \end{bmatrix} = [\frac{3}{8} \quad 0 \quad \frac{1}{4} \quad 0 \quad \frac{3}{8}]$$

Example 4

Calculate T^2 and T^3 when T is the matrix (1) in the random-walk problem.

SOLUTION

$$T = \begin{bmatrix} 1 & 0 & 0 & 0 & 0 & 0 \\ q & 0 & p & 0 & 0 & 0 \\ 0 & q & 0 & p & 0 & 0 \\ 0 & 0 & q & 0 & p & 0 \\ 0 & 0 & 0 & q & 0 & p \\ 0 & 0 & 0 & 0 & 0 & 1 \end{bmatrix}$$

$$T^2 = \begin{bmatrix} 1 & 0 & 0 & 0 & 0 & 0 \\ q & 0 & p & 0 & 0 & 0 \\ 0 & q & 0 & p & 0 & 0 \\ 0 & 0 & q & 0 & p & 0 \\ 0 & 0 & 0 & q & 0 & p \\ 0 & 0 & 0 & 0 & 0 & 1 \end{bmatrix} \cdot \begin{bmatrix} 1 & 0 & 0 & 0 & 0 & 0 \\ q & 0 & p & 0 & 0 & 0 \\ 0 & q & 0 & p & 0 & 0 \\ 0 & 0 & q & 0 & p & 0 \\ 0 & 0 & 0 & q & 0 & p \\ 0 & 0 & 0 & 0 & 0 & 1 \end{bmatrix}$$

$$= \begin{bmatrix} 1 & 0 & 0 & 0 & 0 & 0 \\ q & pq & 0 & p^2 & 0 & 0 \\ q^2 & 0 & 2pq & 0 & p^2 & 0 \\ 0 & q^2 & 0 & 2pq & 0 & p^2 \\ 0 & 0 & q^2 & 0 & pq & p \\ 0 & 0 & 0 & 0 & 0 & 1 \end{bmatrix}$$

$$T^3 = \begin{bmatrix} 1 & 0 & 0 & 0 & 0 & 0 \\ q & pq & 0 & p^2 & 0 & 0 \\ q^2 & 0 & 2pq & 0 & p^2 & 0 \\ 0 & q^2 & 0 & 2pq & 0 & p^2 \\ 0 & 0 & q^2 & 0 & pq & p \\ 0 & 0 & 0 & 0 & 0 & 1 \end{bmatrix} \cdot \begin{bmatrix} 1 & 0 & 0 & 0 & 0 & 0 \\ q & 0 & p & 0 & 0 & 0 \\ 0 & q & 0 & p & 0 & 0 \\ 0 & 0 & q & 0 & p & 0 \\ 0 & 0 & 0 & q & 0 & p \\ 0 & 0 & 0 & 0 & 0 & 1 \end{bmatrix}$$

$$= \begin{bmatrix} 1 & 0 & 0 & 0 & 0 & 0 \\ q + pq^2 & 0 & 2p^2q & 0 & p^3 & 0 \\ q^2 & 2pq^2 & 0 & 3p^2q & 0 & p^3 \\ q^3 & 0 & 3pq^2 & 0 & 2p^2q & p^2 \\ 0 & q^3 & 0 & 2pq^2 & 0 & p + p^2q \\ 0 & 0 & 0 & 0 & 0 & 1 \end{bmatrix}$$

EXERCISES

1. Which of the following are probability vectors?

(a) $[0 \quad -1 \quad 1]$

(b) $[\frac{1}{6} \quad \frac{2}{6} \quad \frac{3}{6}]$

(c) $[\frac{1}{10} \quad 0 \quad \frac{9}{10}]$

(d) $[0 \quad 0 \quad 0 \quad 1]$

(e) $[1 \quad 0 \quad \frac{1}{4} \quad -\frac{1}{4}]$

(f) $[0 \quad \frac{1}{3} \quad 0 \quad \frac{2}{3}]$

(g) $[\frac{1}{4} \quad \frac{1}{3} \quad 0 \quad \frac{1}{2}]$

(h) $[\frac{1}{3} \quad \frac{1}{3} \quad \frac{1}{3} \quad 0]$

2. Use the formula

$$p^{(k)} = p^{(k-1)} \begin{bmatrix} \frac{1}{2} & \frac{1}{2} \\ 1 & 0 \end{bmatrix}$$

to find $p^{(k)} = [p_1{}^{(k)} \quad p_2{}^{(k)}]$ for $k = 1, 2, 3$ when

(a) $p^{(0)} = [0 \quad 1]$
(b) $p^{(0)} = [1 \quad 0]$
(c) $p^{(0)} = [\frac{1}{2} \quad \frac{1}{2}]$
(d) $p^{(0)} = [\frac{3}{4} \quad \frac{1}{4}]$

3. Use the formula

$$p^{(k)} = p^{(k-1)} \begin{bmatrix} \frac{1}{3} & \frac{2}{3} \\ \frac{2}{3} & \frac{1}{3} \end{bmatrix}$$

to find $p^{(k)}$ for $k = 1, 2, 3$, when

(a) $p^{(0)} = [0 \quad 1]$
(b) $p^{(0)} = [\frac{2}{3} \quad \frac{1}{3}]$
(c) $p^{(0)} = [1 \quad 0]$
(d) $p^{(0)} = [\frac{1}{4} \quad \frac{3}{4}]$

4. Use the formula

$$p^{(k)} = p^{(0)} \begin{bmatrix} 1 & 0 \\ \frac{1}{2} & \frac{1}{2} \end{bmatrix}^k$$

to find $p^{(k)}$ for $k = 1, 2, 3$ when

(a) $p^{(0)} = [1 \quad 0]$
(b) $p^{(0)} = [\frac{1}{4} \quad \frac{3}{4}]$
(c) $p^{(0)} = [0 \quad 1]$

Find T^2 and T^3 in Exercises 5–17.

5. $\begin{bmatrix} \frac{1}{2} & \frac{1}{2} \\ \frac{1}{4} & \frac{3}{4} \end{bmatrix}$ 6. $\begin{bmatrix} \frac{1}{3} & \frac{2}{3} \\ 1 & 0 \end{bmatrix}$

7. $\begin{bmatrix} 0 & 1 \\ 1 & 0 \end{bmatrix}$ 8. $\begin{bmatrix} 1 & 0 & 0 \\ \frac{1}{2} & 0 & \frac{1}{2} \\ 0 & 0 & 1 \end{bmatrix}$

9. $\begin{bmatrix} 0 & \frac{1}{3} & \frac{2}{3} \\ 0 & 0 & 1 \\ 0 & 1 & 0 \end{bmatrix}$ 10. $\begin{bmatrix} 0 & 1 & 0 \\ 1 & 0 & 0 \\ 0 & 0 & 1 \end{bmatrix}$

11. $\begin{bmatrix} \frac{1}{2} & \frac{1}{2} & 0 \\ \frac{1}{2} & 0 & \frac{1}{2} \\ 0 & \frac{1}{2} & \frac{1}{2} \end{bmatrix}$
  12. $\begin{bmatrix} 0 & \frac{1}{2} & \frac{1}{2} \\ \frac{1}{2} & 0 & \frac{1}{2} \\ \frac{1}{2} & \frac{1}{2} & 0 \end{bmatrix}$

13. $\begin{bmatrix} \frac{1}{4} & \frac{1}{4} & \frac{1}{2} \\ \frac{1}{4} & \frac{1}{2} & \frac{1}{4} \\ \frac{1}{2} & \frac{1}{4} & \frac{1}{4} \end{bmatrix}$
  14. $\begin{bmatrix} \frac{1}{2} & \frac{1}{4} & \frac{1}{4} \\ \frac{1}{4} & \frac{1}{2} & \frac{1}{4} \\ \frac{1}{4} & \frac{1}{4} & \frac{1}{2} \end{bmatrix}$

15. $\begin{bmatrix} 1 & 0 & 0 & 0 \\ \frac{1}{2} & 0 & \frac{1}{2} & 0 \\ 0 & \frac{1}{2} & 0 & \frac{1}{2} \\ 0 & 0 & 0 & 1 \end{bmatrix}$
  16. $\begin{bmatrix} 0 & 1 & 0 & 0 \\ \frac{1}{2} & 0 & \frac{1}{2} & 0 \\ 0 & \frac{1}{2} & 0 & \frac{1}{2} \\ 0 & 0 & 1 & 0 \end{bmatrix}$

17. $\begin{bmatrix} 0 & 0 & \frac{1}{2} & \frac{1}{2} \\ \frac{1}{2} & \frac{1}{2} & 0 & 0 \\ 0 & 0 & 1 & 0 \\ 0 & 0 & 0 & 1 \end{bmatrix}$

In Exercises 18–20 find the k-step probability vectors

$$p^{(k)} = [p_1^{(k)} \quad p_2^{(k)} \quad p_3^{(k)}] \quad \text{for} \quad k = 1, 2, 3$$

for the given matrix T and initial probability vectors $p^{(0)}$.

18. T is as in Exercise 12, and

 (a) $p^{(0)} = [0 \quad 1 \quad 0]$
 (b) $p^{(0)} = [0 \quad 0 \quad 1]$

19. T is as in Exercise 9, and

 (a) $p^{(0)} = [0 \quad 1 \quad 0]$
 (b) $p^{(0)} = [1 \quad 0 \quad 0]$

20. T is as in Exercise 10, and

 (a) $p^{(0)} = [\frac{1}{3} \quad \frac{1}{3} \quad \frac{1}{3}]$
 (b) $p^{(0)} = [\frac{1}{2} \quad 0 \quad \frac{1}{2}]$

7.3 REGULAR MARKOV CHAINS
In Section 7.1 we introduced the concept of *probability vector* and we
learned how to calculate transition matrices and probability vectors
for successive transitions of a Markov chain. In this section we shall

examine an important class of transition matrices called *regular* matrices. This class is introduced by means of the following example.

Example 1

Because of exposure to advertising, a person acquired the habit of buying each day on his way to work a pack of one of two brands of chewing gum. There is a probability of $\frac{3}{4}$ that he will buy on any given day the same brand he bought last time. If the first pack was brand A, what is the probability that the person's kth purchase will also be brand A?

SOLUTION

Let the other brand be B. Then the chain of successive gum purchases can be described by means of two states. Namely, we shall say that the process is in state A at the time k if the kth purchase was brand A; otherwise, we shall say that the process is in state B. The Markov chain can therefore be described by means of the transition matrix

$$
\begin{array}{cc} & A \ \ B \end{array}
$$
$$
\begin{array}{c} A \\ B \end{array}
\begin{bmatrix} \frac{3}{4} & \frac{1}{4} \\ \frac{1}{4} & \frac{3}{4} \end{bmatrix} = T
$$

The initial state of the process is given by means of the probability vector

$$
p^{(0)} = [1 \quad 0]
$$

and at time k the k-step probability vector is

$$
p^{(k)} = [p_1^{(k)} \quad p_2^{(k)}]
$$

Here $p_1^{(k)}$ is the probability of being in state A at time k, and $p_2^{(k)}$ is the probability of being in state B at time k.

To calculate $p^{(k)}$ we recall the iteration formula

$$
p^{(k)} = p^{(k-1)} \begin{bmatrix} \frac{3}{4} & \frac{1}{4} \\ \frac{1}{4} & \frac{3}{4} \end{bmatrix} \tag{9}
$$

(See Formula (6) in Section 7.2); it gives the following

$$
p^{(1)} = [1 \quad 0] \cdot \begin{bmatrix} \frac{3}{4} & \frac{1}{4} \\ \frac{1}{4} & \frac{3}{4} \end{bmatrix} = [\tfrac{3}{4} \quad \tfrac{1}{4}] = [(\tfrac{1}{2} + \tfrac{1}{4}) \quad (\tfrac{1}{2} - \tfrac{1}{4})]
$$

$$p^{(2)} = \begin{bmatrix} \frac{3}{4} & \frac{1}{4} \end{bmatrix} \cdot \begin{bmatrix} \frac{3}{4} & \frac{1}{4} \\ \frac{1}{4} & \frac{3}{4} \end{bmatrix} = \begin{bmatrix} \frac{5}{8} & \frac{3}{8} \end{bmatrix} = \begin{bmatrix} (\frac{1}{2} + \frac{1}{8}) & (\frac{1}{2} - \frac{1}{8}) \end{bmatrix}$$

$$p^{(3)} = \begin{bmatrix} \frac{5}{8} & \frac{3}{8} \end{bmatrix} \cdot \begin{bmatrix} \frac{3}{4} & \frac{1}{4} \\ \frac{1}{4} & \frac{3}{4} \end{bmatrix} = \begin{bmatrix} \frac{9}{16} & \frac{7}{16} \end{bmatrix} = \begin{bmatrix} (\frac{1}{2} + \frac{1}{16}) & (\frac{1}{2} - \frac{1}{16}) \end{bmatrix}$$

$$p^{(4)} = \begin{bmatrix} \frac{9}{16} & \frac{7}{16} \end{bmatrix} \cdot \begin{bmatrix} \frac{3}{4} & \frac{1}{4} \\ \frac{1}{4} & \frac{3}{4} \end{bmatrix} = \begin{bmatrix} \frac{17}{32} & \frac{15}{32} \end{bmatrix} = \begin{bmatrix} (\frac{1}{2} + \frac{1}{32}) & (\frac{1}{2} - \frac{1}{32}) \end{bmatrix}$$

These calculations indicate that, in general,

$$p^{(k)} = \left[\left(\frac{1}{2} + \frac{1}{2^{k+1}} \right) \quad \left(\frac{1}{2} - \frac{1}{2^{k+1}} \right) \right]$$

and we are led to the interesting discovery that as k becomes bigger and bigger, the numbers $p_1^{(k)}$ and $p_2^{(k)}$ come closer and closer to $\frac{1}{2}$. We say that the probability vectors $p^{(k)}$ *approach* the probability vector $u = \begin{bmatrix} \frac{1}{2} & \frac{1}{2} \end{bmatrix}$ as k becomes bigger and bigger (or as k *approaches infinity*). This discovery has the following interpretation: In the long run (i,e., for large values of k), the person can be expected to purchase brand A and brand B with about equal probability $\frac{1}{2}$. The initial state of the process is thus seen to be unimportant here.

It is seen intuitively that both $p^{(k)}$ and $p^{(k-1)}$ approach the vector $u = \begin{bmatrix} \frac{1}{2} & \frac{1}{2} \end{bmatrix}$ as k becomes bigger. From Equation (9) we therefore deduce the equation

$$\begin{bmatrix} \frac{1}{2} & \frac{1}{2} \end{bmatrix} = \begin{bmatrix} \frac{1}{2} & \frac{1}{2} \end{bmatrix} \begin{bmatrix} \frac{3}{4} & \frac{1}{4} \\ \frac{1}{4} & \frac{3}{4} \end{bmatrix}$$

A simple calculation shows that this equation is true. The vector u is said to be a fixed point of the matrix.

DEFINITION 1

A probability vector u is a *fixed point* of the matrix T if

$$uT = u$$

There is another important fact that we can learn from the preceding problem. Namely, from Section 7.2 we have the formula

$$p^{(k)} = \begin{bmatrix} 1 & 0 \end{bmatrix} \begin{bmatrix} \frac{3}{4} & \frac{1}{4} \\ \frac{1}{4} & \frac{3}{4} \end{bmatrix}^k \tag{10}$$

which in abbreviated form is $p^{(k)} = p^{(0)} T^k$.

We are now curious about the matrices T^k: Since the vectors $p^{(k)}$ approach the fixed vector u, is it not necessary for the matrices T^k to approach in the same sense some fixed matrix U? Specifically, substituting u for $p^{(k)}$ in Equation (10) gives the approximate equality

$$[\tfrac{1}{2} \quad \tfrac{1}{2}] \approx [1 \quad 0] \begin{bmatrix} \tfrac{3}{4} & \tfrac{1}{4} \\ \tfrac{1}{4} & \tfrac{3}{4} \end{bmatrix}^k$$

and it appears intuitively clear that the right side should approach the left side as k becomes bigger and bigger. A direct calculation gives the following:

$$T^2 = \begin{bmatrix} \tfrac{3}{4} & \tfrac{1}{4} \\ \tfrac{1}{4} & \tfrac{3}{4} \end{bmatrix} \cdot \begin{bmatrix} \tfrac{3}{4} & \tfrac{1}{4} \\ \tfrac{1}{4} & \tfrac{3}{4} \end{bmatrix} = \begin{bmatrix} \tfrac{5}{8} & \tfrac{3}{8} \\ \tfrac{3}{8} & \tfrac{5}{8} \end{bmatrix}$$

$$T^3 = \begin{bmatrix} \tfrac{5}{8} & \tfrac{3}{8} \\ \tfrac{3}{8} & \tfrac{5}{8} \end{bmatrix} \cdot \begin{bmatrix} \tfrac{3}{4} & \tfrac{1}{4} \\ \tfrac{1}{4} & \tfrac{3}{4} \end{bmatrix} = \begin{bmatrix} \tfrac{9}{16} & \tfrac{7}{16} \\ \tfrac{7}{16} & \tfrac{9}{16} \end{bmatrix}$$

$$T^4 = \begin{bmatrix} \tfrac{9}{16} & \tfrac{7}{16} \\ \tfrac{7}{16} & \tfrac{9}{16} \end{bmatrix} \cdot \begin{bmatrix} \tfrac{3}{4} & \tfrac{1}{4} \\ \tfrac{1}{4} & \tfrac{3}{4} \end{bmatrix} = \begin{bmatrix} \tfrac{17}{32} & \tfrac{15}{32} \\ \tfrac{15}{32} & \tfrac{17}{32} \end{bmatrix}$$

Comparing the matrices on the right with the calculations for $p^{(k)}$, we find that each entry of the matrices T^k approaches $\tfrac{1}{2}$ as k increases, and we are led to conclude that the matrices T^k approach the fixed matrix

$$U = \begin{bmatrix} \tfrac{1}{2} & \tfrac{1}{2} \\ \tfrac{1}{2} & \tfrac{1}{2} \end{bmatrix}$$

We see that indeed

$$[1 \quad 0]U = [1 \quad 0] \begin{bmatrix} \tfrac{1}{2} & \tfrac{1}{2} \\ \tfrac{1}{2} & \tfrac{1}{2} \end{bmatrix} = [\tfrac{1}{2} \quad \tfrac{1}{2}] = u$$

The matrix T belongs to the class of matrices called *regular*.

DEFINITION 2

A transition matrix $T = [p_{ij}]$ is said to be *regular* if there is a power k for which the elements of T^k are all positive.

Example 2
The matrix

$$T = \begin{bmatrix} 0 & 1 \\ \tfrac{1}{5} & \tfrac{4}{5} \end{bmatrix}$$

is regular, since

$$T^2 = \begin{bmatrix} 0 & 1 \\ \frac{1}{5} & \frac{4}{5} \end{bmatrix} \cdot \begin{bmatrix} 0 & 1 \\ \frac{1}{5} & \frac{4}{5} \end{bmatrix} = \begin{bmatrix} \frac{1}{5} & \frac{4}{5} \\ \frac{4}{25} & \frac{16}{25} \end{bmatrix}$$

The matrix

$$T = \begin{bmatrix} 1 & 0 \\ 1 & 0 \end{bmatrix}$$

is not regular, since

$$T^2 = \begin{bmatrix} 1 & 0 \\ 1 & 0 \end{bmatrix} \cdot \begin{bmatrix} 1 & 0 \\ 1 & 0 \end{bmatrix} = \begin{bmatrix} 1 & 0 \\ 1 & 0 \end{bmatrix} = T$$

and we see that for no power k does T^k have all positive elements.

When a Markov process has a regular transition matrix, we shall call the process *regular*.

DEFINITION 3
A Markov process is said to be *regular* if and only if it has a regular transition matrix.

The significance of being regular is that in such a process it is possible to be in *any* state after a finite number of steps (transitions). The properties of the matrix T in Example 1 are characteristic of regular matrices in general. We summarize these properties in the following theorem.

THEOREM 1
Let T be a regular matrix. Then

(1) The matrices T^k approach a fixed matrix U.
(2) The rows of U are one and the same probability vector.
(3) All the elements of U are positive.

Example 3
Consider the matrix

$$T = \begin{bmatrix} \frac{1}{2} & \frac{1}{2} & 0 \\ \frac{1}{2} & 0 & \frac{1}{2} \\ 0 & \frac{1}{2} & \frac{1}{2} \end{bmatrix}$$

Since

$$T^2 = \begin{bmatrix} \frac{1}{2} & \frac{1}{2} & 0 \\ \frac{1}{2} & 0 & \frac{1}{2} \\ 0 & \frac{1}{2} & \frac{1}{2} \end{bmatrix} \cdot \begin{bmatrix} \frac{1}{2} & \frac{1}{2} & 0 \\ \frac{1}{2} & 0 & \frac{1}{2} \\ 0 & \frac{1}{2} & \frac{1}{2} \end{bmatrix} = \begin{bmatrix} \frac{1}{2} & \frac{1}{4} & \frac{1}{4} \\ \frac{1}{4} & \frac{1}{2} & \frac{1}{4} \\ \frac{1}{4} & \frac{1}{4} & \frac{1}{2} \end{bmatrix}$$

we see that T is regular. What matrix U do the matrices T^k approach as k increases? To decide this, we calculate a few more powers of T:

$$T^3 = \begin{bmatrix} \frac{3}{8} & \frac{3}{8} & \frac{1}{4} \\ \frac{3}{8} & \frac{1}{4} & \frac{3}{8} \\ \frac{1}{4} & \frac{3}{8} & \frac{3}{8} \end{bmatrix}$$

$$T^4 = \begin{bmatrix} \frac{3}{8} & \frac{5}{16} & \frac{5}{16} \\ \frac{5}{16} & \frac{3}{8} & \frac{5}{16} \\ \frac{5}{16} & \frac{5}{16} & \frac{3}{8} \end{bmatrix}$$

$$T^5 = \begin{bmatrix} \frac{11}{32} & \frac{11}{32} & \frac{5}{16} \\ \frac{11}{32} & \frac{5}{16} & \frac{11}{32} \\ \frac{5}{16} & \frac{11}{32} & \frac{11}{32} \end{bmatrix}$$

How can we discover a pattern which will tell us what the elements of T^k approach as k becomes big? There is really no easy answer, and the reader will have to inspect the entries in the matrices T, T^2, T^3, and so on, very carefully. In this case, for instance, we see that the entries in any of the matrices we computed is "approximately" $\frac{1}{3}$. More specifically, we get $\frac{1}{3}$ if we increase or decrease the denominators by 1. Thus, in T^4 we see that

$$\frac{3}{8+1} = \frac{1}{3} \quad \text{and} \quad \frac{5}{16-1} = \frac{1}{3}$$

In T^5 we see that

$$\frac{11}{32+1} = \frac{1}{3} \quad \text{and} \quad \frac{5}{16-1} = \frac{1}{3}$$

Once we discover this fact, we observe that the 1 we add or subtract becomes less significant for large k because the numerators and denominators of T^k are large. We thus conclude that the elements of T^k approach $\frac{1}{3}$ as k increases. Hence, the matrices T^k approach the matrix

$$U = \begin{bmatrix} \frac{1}{3} & \frac{1}{3} & \frac{1}{3} \\ \frac{1}{3} & \frac{1}{3} & \frac{1}{3} \\ \frac{1}{3} & \frac{1}{3} & \frac{1}{3} \end{bmatrix}$$

Another property of regular matrices is this:

THEOREM 2

Let T be a regular matrix and let the matrices T^k approach the fixed matrix U. If p is any probability vector, then the probability vectors pT^k approach the unique fixed point u of T.

This fact was already used in Example 1.

Example 4

In Example 1, suppose that the probability that the person will buy on any given day the brand of the last purchase is $\frac{3}{4}$ if it was brand A, and $\frac{1}{2}$ if it was brand B. If his first purchase was brand A, what is the probability that it will be brand A in the long run?

SOLUTION

This Markov process is described by the transition matrix

$$
\begin{array}{cc} & A \quad B \end{array}
$$
$$
\begin{array}{c} A \\ B \end{array}
\begin{bmatrix} \frac{3}{4} & \frac{1}{4} \\ \frac{1}{2} & \frac{1}{2} \end{bmatrix} = S
$$

By the preceding discussion, there is a probability vector $u = [u_1 \quad u_2]$ such that

$$
[u_1 \quad u_2] = [u_1 \quad u_2] \cdot \begin{bmatrix} \frac{3}{4} & \frac{1}{4} \\ \frac{1}{2} & \frac{1}{2} \end{bmatrix}
$$
$$
= [\tfrac{3}{4}u_1 + \tfrac{1}{2}u_2 \quad \tfrac{1}{4}u_1 + \tfrac{1}{2}u_2]
$$

This gives rise to the system of equations

$$
\begin{aligned}
\tfrac{3}{4}u_1 + \tfrac{1}{2}u_2 &= u_1 \\
\tfrac{1}{4}u_1 + \tfrac{1}{2}u_2 &= u_2 \\
u_1 + u_2 &= 1
\end{aligned}
$$

The third equation in this system is due to the fact that $[u_1 \quad u_2]$ is a probability vector. Each of the two top equations can be simplified to $u_1 = 2u_2$, and this leaves us with the system

$$
\begin{aligned}
u_1 - 2u_2 &= 0 \\
u_1 + u_2 &= 1
\end{aligned}
$$

Solving this system of linear equations gives

$$u_1 = \tfrac{2}{3} \qquad u_2 = \tfrac{1}{3}$$

To check our answer, we perform the following calculation:

$$[\tfrac{2}{3} \quad \tfrac{1}{3}] \cdot \begin{bmatrix} \tfrac{3}{4} & \tfrac{1}{4} \\ \tfrac{1}{2} & \tfrac{1}{2} \end{bmatrix} = [\tfrac{6}{12} + \tfrac{1}{6} \quad \tfrac{2}{12} + \tfrac{1}{6}] = [\tfrac{2}{3} \quad \tfrac{1}{3}]$$

Hence, $u = [\tfrac{2}{3} \quad \tfrac{1}{3}]$ is the fixed point of the matrix S. We have thus found that, in the long run, the probability is about $\tfrac{2}{3}$ that the person will purchase brand A on any given day.

EXERCISES

1. Decide which of the following matrices are regular.

(a) $\begin{bmatrix} 0 & 1 \\ \tfrac{1}{2} & \tfrac{1}{2} \end{bmatrix}$ (b) $\begin{bmatrix} 1 & 0 \\ \tfrac{1}{2} & \tfrac{1}{2} \end{bmatrix}$

(c) $\begin{bmatrix} \tfrac{1}{2} & \tfrac{1}{2} \\ 1 & 0 \end{bmatrix}$ (d) $\begin{bmatrix} \tfrac{1}{2} & \tfrac{1}{2} \\ 0 & 1 \end{bmatrix}$

(e) $\begin{bmatrix} 0 & 1 \\ 1 & 0 \end{bmatrix}$ (f) $\begin{bmatrix} \tfrac{1}{2} & \tfrac{1}{2} & 0 \\ \tfrac{1}{2} & 0 & \tfrac{1}{2} \\ 0 & 1 & 0 \end{bmatrix}$

(g) $\begin{bmatrix} 0 & 1 & 0 \\ 1 & 0 & 0 \\ 0 & 0 & 1 \end{bmatrix}$ (h) $\begin{bmatrix} \tfrac{1}{4} & 0 & \tfrac{3}{4} \\ 0 & 1 & 0 \\ 0 & \tfrac{4}{5} & \tfrac{1}{5} \end{bmatrix}$

2. Find the unique fixed point (probability vector) of each of the following matrices.

(a) $\begin{bmatrix} \tfrac{4}{5} & \tfrac{1}{5} \\ \tfrac{1}{2} & \tfrac{1}{2} \end{bmatrix}$

(b) $\begin{bmatrix} \tfrac{4}{5} & \tfrac{1}{5} \\ \tfrac{1}{6} & \tfrac{5}{6} \end{bmatrix}$

(c) $\begin{bmatrix} 1-a & a \\ a & 1-a \end{bmatrix}$ $(0 < a < 1)$

(d) $\begin{bmatrix} 1-a & a \\ b & 1-b \end{bmatrix}$ $(0 < a < 1 \quad \text{and} \quad 0 < b < 1)$

(e) $\begin{bmatrix} \frac{3}{4} & \frac{1}{4} & 0 \\ 0 & \frac{2}{3} & \frac{1}{3} \\ \frac{1}{4} & \frac{1}{4} & \frac{1}{2} \end{bmatrix}$

(f) $\begin{bmatrix} \frac{1}{2} & \frac{1}{4} & \frac{1}{4} \\ \frac{1}{2} & 0 & \frac{1}{2} \\ 0 & 1 & 0 \end{bmatrix}$

3. An incumbent politician tells a friend that he will run for reelection. The friend relays the news to another friend, and so on: It is assumed that the news is relayed each time to a new person. Suppose that there is a probability $\frac{1}{2}$ that any given person will reverse the news before he relays it to the next person. In the long run, what is the probability that a person to receive the news will be told that the incumbent will run for reelection?

4. What is the answer to Exercise 3 if the probability for reversing the news is $\frac{1}{10}$?

5. In a defective vending machine, if it dispenses a wrong item, there is a probability of $\frac{24}{25}$ that the next dispensed item will be correct; if the machine dispenses a correct item, there is a probability of $\frac{1}{10}$ that the next item will be wrong. In the long run, what is the probability that the machine will dispense the correct item?

6. A store carries three brands of toothpaste: A, B, and C. A survey shows that the following behavior can be expected from any given female customer:

    If she last bought brand A, she will buy brand A or B next time, each with probability $\frac{1}{2}$.
    If she last bought brand B, she will buy brands B or C next time, each with probability $\frac{1}{2}$.
    If she last bought brand C, there is a probability of $\frac{1}{4}$ that she will buy brand C next time, and a probability of $\frac{3}{4}$ that she will buy brand A.

    (a) Show that this Markov process is regular.
    (b) In the long run, what is the probability that a customer will buy brand C toothpaste?

7. An incumbent politician tells a friend that he will run for reelection. This news is relayed from one person to another, but it

turned out that any person down the line could have been told one of three things:

(a)   The incumbent will run.
(b)   The incumbent will not run.
(c)   Maybe the incumbent will run.

For any person down the line, the following is true:

He will relay the news as he received it with probability $\frac{1}{2}$.
If he was told (a) or (b), he will reverse the news before relaying it, with probability $\frac{1}{4}$.
If he was told (c), he will change it to (b) before relaying it, with probability $\frac{1}{2}$.

(a)   Show that this Markov process is regular.
(b)   In the long run, what is the probability that a person will be told that the incumbent will run for reelection?

8.   From past experience, a car manufacturer has the following expectations:

A compact-car owner will replace it with another compact car or an intermediate car, each with probability $\frac{1}{2}$.
An intermediate-car owner will replace it with another intermediate with probability 1.
A full-size car owner will replace it with an intermediate car with probability $\frac{3}{4}$, or another full-size car with probability $\frac{1}{4}$.

(a)   Is this Markov process regular?
(b)   What is the probability that the fourth car of a compact-car owner will be compact?
(c)   What is the probability that the fourth car of a full-size car owner will be intermediate?

7.4   ABSORBING MARKOV CHAINS

In the random-walk problem introduced in Section 7.1, we noted that two states, a_1 and a_6, were terminal states in the sense that once the process reached one of them, it remained there. We called these states *absorbing barriers*. The type of process with an absorbing barrier is very different from the type of process we called *regular* in Section 7.3. To fix ideas, we start with a formal definition.

DEFINITION 1

A state in a Markov chain is an *absorbing state* if, once the process reaches it, it remains there. Otherwise, the state is *nonabsorbing.*

Thus, if a_i is an absorbing state, then its k-step transition probabilities

$$p_{ii}^{(0)}, \; p_{ii}^{(1)}, \; p_{ii}^{(2)}, \; \ldots, \text{ and so on}$$

are all equal to 1.

Example 1

Consider states a_1, a_2, a_3 with transition matrices

$$
\begin{array}{c}
 \\
a_1 \\
a_2 \\
a_3
\end{array}
\begin{array}{ccc}
a_1 & a_2 & a_3 \\
\end{array}
\left[\begin{array}{ccc}
\frac{1}{3} & \frac{1}{3} & \frac{1}{3} \\
0 & 1 & 0 \\
\frac{1}{2} & 0 & \frac{1}{2}
\end{array}\right] = T
\qquad
\begin{array}{c}
 \\
a_1 \\
a_2 \\
a_3
\end{array}
\begin{array}{ccc}
a_1 & a_2 & a_3 \\
\end{array}
\left[\begin{array}{ccc}
\frac{1}{3} & \frac{1}{3} & \frac{1}{3} \\
0 & 1 & 0 \\
\frac{1}{2} & \frac{1}{2} & 0
\end{array}\right] = S
$$

The state a_2 is an *absorbing* state in either process. We note that in the process specified by the transition matrix S, the state a_2 can be reached from any state. In the other process, a_2 cannot be reached from state a_3, but a_2 can be reached from any state after one transition, since

$$
T^2 =
\left[\begin{array}{ccc}
\frac{1}{3} & \frac{1}{3} & \frac{1}{3} \\
0 & 1 & 0 \\
\frac{1}{2} & 0 & \frac{1}{2}
\end{array}\right]
\cdot
\left[\begin{array}{ccc}
\frac{1}{3} & \frac{1}{3} & \frac{1}{3} \\
0 & 1 & 0 \\
\frac{1}{2} & 0 & \frac{1}{2}
\end{array}\right]
=
\left[\begin{array}{ccc}
\frac{5}{18} & \frac{4}{9} & \frac{5}{18} \\
0 & 1 & 0 \\
\frac{5}{12} & \frac{1}{6} & \frac{5}{12}
\end{array}\right]
$$

In either case we say that the process is absorbing.

DEFINITION 2

A Markov chain is *absorbing* if

(1)   It has at least one absorbing state.
(2)   It is possible to reach an absorbing state from each non-absorbing state (possibly in more than one step).

When an absorbing process reaches an absorbing state, it is said to be *absorbed.* Absorbing Markov chains have a number of important features which will now be discussed without proof. For definiteness, we shall use a specific example.

Consider the gambler's-ruin transition matrix from Section 7.1. For $p = q = \frac{1}{2}$, this matrix becomes

$$
\begin{array}{c}
 & \begin{array}{ccccc} a_1 & a_2 & a_3 & a_4 & a_5 \end{array} \\
\begin{array}{c} a_1 \\ a_2 \\ a_3 \\ a_4 \\ a_5 \end{array} &
\left[\begin{array}{ccccc}
1 & 0 & 0 & 0 & 0 \\
\frac{1}{2} & 0 & \frac{1}{2} & 0 & 0 \\
0 & \frac{1}{2} & 0 & \frac{1}{2} & 0 \\
0 & 0 & \frac{1}{2} & 0 & \frac{1}{2} \\
0 & 0 & 0 & 0 & 1
\end{array}\right] = G
\end{array}
$$

In this system, the states a_1 and a_5 are absorbing states, and a_2, a_3, and a_4 are nonabsorbing states. Suppose we play the coin-tossing game as described in the gambler's-ruin problem. The following questions are then natural ones:

(1) What is the average number of times the game is expected to be in each of the nonabsorbing states?

(2) Will the process necessarily be absorbed? If so, how long, on the average, do we expect to have to play before this happens?

(3) What is your probability of winning? What is mine?

To answer question (1), we begin by rearranging the order of the states so that the 1's in the matrix G will be in the upper left-hand corner: We write G in the equivalent form

$$
\begin{array}{c}
 & \begin{array}{ccccc} a_1 & a_5 & a_2 & a_3 & a_4 \end{array} \\
\begin{array}{c} a_1 \\ a_5 \\ a_2 \\ a_3 \\ a_4 \end{array} &
\left[\begin{array}{cc|ccc}
1 & 0 & 0 & 0 & 0 \\
0 & 1 & 0 & 0 & 0 \\ \hline
\frac{1}{2} & 0 & 0 & \frac{1}{2} & 0 \\
0 & 0 & \frac{1}{2} & 0 & \frac{1}{2} \\
0 & \frac{1}{2} & 0 & \frac{1}{2} & 0
\end{array}\right] = G
\end{array}
\qquad (11)
$$

The matrix G is now said to be in *standard form*. It is partitioned into four submatrices

$$
G = \left[\begin{array}{c|c} I & 0 \\ \hline F & H \end{array}\right]
\qquad (12)
$$

where

$$
I = \begin{bmatrix} 1 & 0 \\ 0 & 1 \end{bmatrix}
\qquad
0 = \begin{bmatrix} 0 & 0 & 0 \\ 0 & 0 & 0 \end{bmatrix}
$$

$$F = \begin{bmatrix} \frac{1}{2} & 0 \\ 0 & 0 \\ 0 & \frac{1}{2} \end{bmatrix} \qquad H = \begin{bmatrix} 0 & \frac{1}{2} & 0 \\ \frac{1}{2} & 0 & \frac{1}{2} \\ 0 & \frac{1}{2} & 0 \end{bmatrix}$$

Observe that I is the identity matrix, 0 is the zero matrix, and H is the transition matrix of the three nonabsorbing states. The matrix F will be used later. To see how G changes under transitions, we calculate G^2, but we do it schematically, using (12):

$$G^2 = \left[\begin{array}{c|c} I & 0 \\ \hline F & H \end{array} \right] \cdot \left[\begin{array}{c|c} I & 0 \\ \hline F & H \end{array} \right]$$

$$= \left[\begin{array}{c|c} I + 0 \cdot F & I \cdot 0 + 0 \cdot H \\ \hline F \cdot I + H \cdot F & F \cdot 0 + H^2 \end{array} \right]$$

We see that all the matrix products and sums are defined, since the dimensions in each block match. The matrix $F \cdot I + H \cdot F$ is of no interest, and we designate it with an asterisk. The matrix G^2 can thus be written as

$$G^2 = \left[\begin{array}{c|c} I & 0 \\ \hline * & H^2 \end{array} \right]$$

and it is seen intuitively that, for $k = 1, 2, 3, \ldots$,

$$G^k = \left[\begin{array}{c|c} I & 0 \\ \hline * & H^k \end{array} \right] \tag{13}$$

Here $*$ stands for different submatrices for different k.

Example 2
Starting with state a_3, what is the probability of being in state a_4 after five steps?

SOLUTION
We have the matrix

$$\begin{array}{c} \\ a_2 \\ a_3 \\ a_4 \end{array} \begin{array}{ccc} a_2 & a_3 & a_4 \\ \begin{bmatrix} 0 & \frac{1}{2} & 0 \\ \frac{1}{2} & 0 & \frac{1}{2} \\ 0 & \frac{1}{2} & 0 \end{bmatrix} \end{array} = H$$

A direct calculation gives

$$H^2 = \begin{bmatrix} \frac{1}{4} & 0 & \frac{1}{4} \\ 0 & \frac{1}{2} & 0 \\ \frac{1}{4} & 0 & \frac{1}{4} \end{bmatrix} \qquad H^3 = \begin{bmatrix} 0 & \frac{1}{4} & 0 \\ \frac{1}{4} & 0 & \frac{1}{4} \\ 0 & \frac{1}{4} & 0 \end{bmatrix}$$

$$H^4 = \begin{bmatrix} \frac{1}{8} & 0 & \frac{1}{8} \\ 0 & \frac{1}{4} & 0 \\ \frac{1}{8} & 0 & \frac{1}{8} \end{bmatrix} \qquad H^5 = \begin{bmatrix} 0 & \frac{1}{8} & 0 \\ \frac{1}{8} & 0 & \frac{1}{8} \\ 0 & \frac{1}{8} & 0 \end{bmatrix}$$

Hence, the probability of being in state a_4 after five steps is $\frac{1}{8}$.

An important conclusion can be drawn from the calculations in Example 2. We see that the elements of H^k become smaller and smaller as k becomes bigger and bigger. Intuitively, the elements of H^k approach zero as k increases. Hence, in the long run, the probability of being in any nonabsorbing state is about zero, and this leads us to the answer for the first part of question (2). We state it as a general theorem.

THEOREM 1
An absorbing Markov chain has probability 1 of being absorbed.

After all, if the probability of being in a nonabsorbing state approaches zero, then the probability of being in an absorbing state must approach 1.

The answer to question (1) must, regrettably, be given in a very formal and unmotivated manner, since the derivation involves too many ideas and techniques which are only peripheral to our discussion here. From the matrix H, we form the matrix

$$I - H = \begin{bmatrix} 1 & 0 & 0 \\ 0 & 1 & 0 \\ 0 & 0 & 1 \end{bmatrix} - \begin{bmatrix} 0 & \frac{1}{2} & 0 \\ \frac{1}{2} & 0 & \frac{1}{2} \\ 0 & \frac{1}{2} & 0 \end{bmatrix} = \begin{bmatrix} 1 & -\frac{1}{2} & 0 \\ -\frac{1}{2} & 1 & -\frac{1}{2} \\ 0 & -\frac{1}{2} & 1 \end{bmatrix}$$

It is shown in matrix theory texts that this matrix always is invertible (see Section 6.4). The inverse,

$$N = (I - H)^{-1}$$

is called the *fundamental matrix* of the absorbing chain. This matrix has the following important interpretation.

THEOREM 2

Let $N = (I - H)^{-1}$ be the fundamental matrix of an absorbing Markov chain. Then the elements of N give the *average* number of times each nonabsorbing state is expected to be reached from a nonabsorbing state.

Thus, this theorem gives us the answer to question (1).

Example 3

Starting in state a_3, what is the average number of times we expect to be in state a_4?

SOLUTION

Using the method of Section 6.4, we can compute the inverse matrix $N = (I - H)^{-1}$, and we find that

$$
\begin{array}{c}
\begin{array}{ccc} a_2 & a_3 & a_4 \end{array} \\
\begin{array}{c} a_2 \\ a_3 \\ a_4 \end{array}
\begin{bmatrix} \frac{3}{2} & 1 & \frac{1}{2} \\ 1 & 2 & 1 \\ \frac{1}{2} & 1 & \frac{3}{2} \end{bmatrix} = N
\end{array} \tag{14}
$$

Hence, if we start from state a_3, the expected average number of times of being in state a_4 is 1.

The answer to question (2) is obtained by adding the row elements of N. From matrix (14) we obtain the following information:

> If the initial step was a_2, then the expected average number of steps before absorption is $\frac{3}{2} + 1 + \frac{1}{2} = 3$.
> If the initial state was a_3, then the expected average number of steps before absorption is $1 + 2 + 1 = 4$.
> If the initial state was a_4, then the expected average number of steps before absorptions is $\frac{1}{2} + 1 + \frac{3}{2} = 3$.
> The same procedure gives the answer to such a question for a general absorbing Markov chain.

SUMMARY

The answer to question (3) is contained in the following summary:

Consider the absorbing Markov chain with absorbing states $a_1, \ldots, a_m$ and nonabsorbing states $a_{m+1}, \ldots, a_n$. Let its transition matrix be T.

Absorbing   Nonabsorbing
states   states

$a_1 \cdots a_m$ $a_{m+1} \cdots a_n$

$$
T = \begin{matrix}
\text{Absorbing} \\
\text{states}
\end{matrix}
\left\{
\begin{matrix}
a_1 \\ \cdot \\ \cdot \\ \cdot \\ a_m \\
\end{matrix}
\right.
\begin{matrix}
a_{m+1} \\ \cdot \\ \cdot \\ \cdot \\ a_n
\end{matrix}
\right.
\begin{bmatrix}
I & 0 \\
\hline
F & H
\end{bmatrix} = T
$$

Put

$$N = (I - H)^{-1}$$

Then

(1)   The elements of N give the *average* number of times each nonabsorbing state is expected to be reached from a nonabsorbing state.

(2)   Starting from state a_i, the expected average number of transitions before absorption is the *sum* of the ith row of N.

(3)   Starting from a nonabsorbing state a_i, let the probability of absorption in state a_j be b_{ij}; let $B = [b_{ij}]$, $i = m + 1$, ..., n, and $j = 1, 2, \ldots, m$. Then

$$B = N \cdot F \tag{15}$$

Example 4
Referring to the gambler's-ruin problem, what is your probability of winning?

SOLUTION

Recall that you win when you are in state a_5. If, by assumption, you start the game in state a_3, we ask for the absorption probability in state a_5 when the Markov process starts from state a_3. By Formula (15) this is given in

$$N \cdot F = \begin{bmatrix} \frac{3}{2} & 1 & \frac{1}{2} \\ 1 & 2 & 1 \\ \frac{1}{2} & 1 & \frac{3}{2} \end{bmatrix} \cdot \begin{bmatrix} \frac{1}{2} & 0 \\ 0 & 0 \\ 0 & \frac{1}{2} \end{bmatrix} = \begin{array}{c} \\ a_2 \\ a_3 \\ a_4 \end{array} \begin{array}{cc} a_1 & a_5 \end{array} \begin{bmatrix} \frac{3}{4} & \frac{1}{4} \\ \frac{1}{2} & \frac{1}{2} \\ \frac{1}{4} & \frac{3}{4} \end{bmatrix}$$

That is, the probability is $\frac{1}{2}$.

Example 5

As a final illustration, consider an absorbing Markov chain with absorbing state a_1 and nonabsorbing states a_2 and a_3. Let the transition matrix be T:

$$T = \begin{array}{c} \\ a_1 \\ a_2 \\ a_3 \end{array} \begin{array}{ccc} a_1 & a_2 & a_3 \end{array} \left[\begin{array}{c|cc} 1 & 0 & 0 \\ \hline \frac{1}{2} & 0 & \frac{1}{2} \\ \frac{1}{4} & \frac{1}{2} & \frac{1}{4} \end{array} \right]$$

Writing this matrix in the standard form

$$T = \left[\begin{array}{c|c} I & 0 \\ \hline F & H \end{array} \right]$$

gives

$$I = [1] \qquad 0 = [0 \quad 0] \qquad F = \begin{bmatrix} \frac{1}{2} \\ \frac{1}{4} \end{bmatrix} \qquad H = \begin{bmatrix} 0 & \frac{1}{2} \\ \frac{1}{2} & \frac{1}{4} \end{bmatrix}$$

A direct calculation gives

$$I - H = \begin{bmatrix} 1 & 0 \\ 0 & 1 \end{bmatrix} - \begin{bmatrix} 0 & \frac{1}{2} \\ \frac{1}{2} & \frac{1}{4} \end{bmatrix} = \begin{bmatrix} 1 & -\frac{1}{2} \\ -\frac{1}{2} & \frac{3}{4} \end{bmatrix}$$

$$N = (I - H)^{-1} = \begin{array}{c} \\ a_2 \\ a_3 \end{array} \begin{array}{cc} a_2 & a_3 \end{array} \begin{bmatrix} \frac{3}{2} & 1 \\ 1 & 2 \end{bmatrix}$$

$$N \cdot F = \begin{bmatrix} \frac{3}{2} & 1 \\ 1 & 2 \end{bmatrix} \cdot \begin{bmatrix} \frac{1}{2} \\ \frac{1}{4} \end{bmatrix} = \begin{array}{c} \\ a_2 \\ a_3 \end{array} \begin{array}{c} a_1 \end{array} \begin{bmatrix} 1 \\ 1 \end{bmatrix}$$

We can now read off the following information:

(1) Starting from state a_2, the average number of times we expect to be in this state is $\frac{3}{2}$, and the average number of times we expect to be in state a_3 is 1, and so on.

(2) Starting from state a_2, the expected average number of steps before absorption is $\frac{3}{2} + 1 = \frac{5}{2}$; starting from state a_3 gives for the answer $1 + 2 = 3$.

(3) The absorption probability starting from state a_2 or a_3 is 1.

EXERCISES

In Exercises 1–11, decide which of the matrices specify absorbing Markov chains, find their absorbing states, and write them in standard form.

1.
$$\begin{array}{c} \\ a_1 \\ a_2 \end{array} \begin{array}{cc} a_1 & a_2 \\ \begin{bmatrix} 0 & 1 \\ \frac{1}{10} & \frac{9}{10} \end{bmatrix} \end{array}$$

2.
$$\begin{array}{c} \\ a_1 \\ a_2 \end{array} \begin{array}{cc} a_1 & a_2 \\ \begin{bmatrix} \frac{1}{5} & \frac{4}{5} \\ \frac{4}{5} & \frac{1}{5} \end{bmatrix} \end{array}$$

3.
$$\begin{array}{c} \\ a_1 \\ a_2 \\ a_3 \end{array} \begin{array}{ccc} a_1 & a_2 & a_3 \\ \begin{bmatrix} 0 & 0 & 1 \\ \frac{1}{3} & \frac{1}{3} & \frac{1}{3} \\ 0 & 0 & 1 \end{bmatrix} \end{array}$$

4.
$$\begin{array}{c} \\ a_1 \\ a_2 \\ a_3 \end{array} \begin{array}{ccc} a_1 & a_2 & a_3 \\ \begin{bmatrix} \frac{1}{6} & \frac{2}{6} & \frac{3}{6} \\ \frac{2}{6} & \frac{3}{6} & \frac{1}{6} \\ 1 & 0 & 0 \end{bmatrix} \end{array}$$

5.
$$\begin{array}{c} \\ a_1 \\ a_2 \\ a_3 \end{array} \begin{array}{ccc} a_1 & a_2 & a_3 \\ \begin{bmatrix} \frac{1}{3} & \frac{2}{3} & 0 \\ 0 & \frac{1}{3} & \frac{2}{3} \\ \frac{1}{3} & 0 & \frac{2}{3} \end{bmatrix} \end{array}$$

6.
$$\begin{array}{c} \\ a_1 \\ a_2 \\ a_3 \end{array} \begin{array}{ccc} a_1 & a_2 & a_3 \\ \begin{bmatrix} \frac{1}{3} & \frac{2}{3} & 0 \\ 0 & 1 & 0 \\ 0 & \frac{1}{3} & \frac{2}{3} \end{bmatrix} \end{array}$$

7.
$$\begin{array}{c} \\ a_1 \\ a_2 \\ a_3 \\ a_4 \end{array} \begin{array}{cccc} a_1 & a_2 & a_3 & a_4 \\ \begin{bmatrix} 0 & 1 & 0 & 0 \\ \frac{1}{2} & 0 & \frac{1}{4} & \frac{1}{4} \\ \frac{1}{4} & \frac{1}{4} & 0 & \frac{1}{2} \\ 0 & 0 & 1 & 0 \end{bmatrix} \end{array}$$

8.
$$\begin{array}{c} \\ a_1 \\ a_2 \\ a_3 \\ a_4 \end{array} \begin{array}{cccc} a_1 & a_2 & a_3 & a_4 \\ \begin{bmatrix} 0 & \frac{1}{2} & \frac{1}{2} & 0 \\ \frac{1}{2} & 0 & 0 & \frac{1}{2} \\ 0 & 1 & 0 & 0 \\ 0 & 0 & 1 & 0 \end{bmatrix} \end{array}$$

9.
$$\begin{array}{c} \\ a_1 \\ a_2 \\ a_3 \\ a_4 \end{array} \begin{array}{cccc} a_1 & a_2 & a_3 & a_4 \\ \begin{bmatrix} \frac{1}{2} & 0 & \frac{1}{2} & 0 \\ 0 & 1 & 0 & 0 \\ 0 & 0 & 1 & 0 \\ \frac{1}{4} & \frac{1}{4} & \frac{1}{4} & \frac{1}{4} \end{bmatrix} \end{array}$$

10.
$$\begin{array}{c} \\ a_1 \\ a_2 \\ a_3 \\ a_4 \\ a_5 \end{array} \begin{array}{ccccc} a_1 & a_2 & a_3 & a_4 & a_5 \\ \begin{bmatrix} 0 & 1 & 0 & 0 & 0 \\ \frac{1}{4} & 0 & \frac{1}{4} & \frac{1}{4} & \frac{1}{4} \\ 0 & 0 & 1 & 0 & 0 \\ \frac{1}{4} & \frac{1}{4} & \frac{1}{4} & 0 & \frac{1}{4} \\ 0 & 0 & 0 & 1 & 0 \end{bmatrix} \end{array}$$

$$
\begin{array}{c}
\begin{array}{ccccc} a_1 & a_2 & a_3 & a_4 & a_5 \end{array} \\
\begin{array}{c} a_1 \\ a_2 \\ \text{11.} \quad a_3 \\ a_4 \\ a_5 \end{array}
\left[
\begin{array}{ccccc}
\frac{2}{7} & \frac{2}{7} & \frac{2}{7} & 0 & \frac{1}{7} \\
\frac{1}{7} & \frac{2}{7} & 0 & \frac{2}{7} & \frac{2}{7} \\
\frac{2}{7} & 0 & \frac{1}{7} & \frac{2}{7} & \frac{2}{7} \\
0 & 0 & 0 & 1 & 0 \\
0 & 0 & 0 & 0 & 1
\end{array}
\right]
\end{array}
$$

12.   Consider the absorbing Markov process:

$$
\begin{array}{c}
\begin{array}{ccc} a_1 & a_2 & a_3 \end{array} \\
\begin{array}{c} a_1 \\ a_2 \\ a_3 \end{array}
\left[
\begin{array}{ccc}
1 & 0 & 0 \\
\frac{1}{3} & \frac{1}{3} & \frac{1}{3} \\
0 & \frac{1}{2} & \frac{1}{2}
\end{array}
\right]
\end{array}
$$

(a)   Starting from state a_2, what is the average number of times we expect to be in state a_3?

(b)   Starting from state a_2, what is the expected average number of steps before absorption?

13.   Consider the absorbing Markov process:

$$
\begin{array}{c}
\begin{array}{cccc} a_1 & a_2 & a_3 & a_4 \end{array} \\
\begin{array}{c} a_1 \\ a_2 \\ a_3 \\ a_4 \end{array}
\left[
\begin{array}{cccc}
1 & 0 & 0 & 0 \\
0 & 1 & 0 & 0 \\
\frac{1}{2} & 0 & \frac{1}{4} & \frac{1}{4} \\
0 & \frac{1}{2} & \frac{1}{4} & \frac{1}{4}
\end{array}
\right]
\end{array}.
$$

(a)   Starting from state a_3, what is the average number of times we expect to be in state a_4?

(b)   Starting from state a_3, what is the expected average number of steps before absorption?

(c)   Starting from state a_4, what is the probability of absorption in state a_1? In state a_2?

ANSWERS
TO
ODD-NUMBERED
EXERCISES

ANSWERS TO ODD-NUMBERED EXERCISES

CHAPTER 0

Section 0.1

1. (a) $=$ (b) $\neq$ (c) $\neq$ (d) $=$ (e) $\neq$
   (f) $\neq$ (g) $=$ (h) $=$

3. -36 5. $\frac{1}{169}$ 7. $\frac{5}{9}$ 9. -1 11. -1

13. $\frac{5}{6} - \frac{1}{7} = \frac{29}{42}$ 15. $\frac{1}{2}(1 + \frac{1}{2} + \frac{1}{4}) = \frac{1}{2} \times \frac{4 + 2 + 1}{4} = \frac{7}{8}$

17. $\dfrac{b}{a + b}$

Section 0.2

1. (a) $\neq$ (b) $=$ (c) $\neq$ (d) $=$ (e) $=$
   (f) $=$ (g) $\neq$ (h) $\neq$

3. $2/3^4$ 5. $2^4/3^6$ 7. $8^3/9^4 = 2^9/3^8$ 9. $1/10^3$

11. $\left(\dfrac{1}{a^3}\right)^{-3} = (a^3)^3 = a^9$ 13. 10^2 15. $3^{3-4+5} = 3^4$

17. $2^4/5^8$ 19. $a^4 b^4 = (ab)^4$

21. $\dfrac{1 - \dfrac{1}{2^{11}}}{1 - \dfrac{1}{2}} = 2\left(1 - \dfrac{1}{2^{11}}\right) = 2 - \dfrac{1}{2^{10}}$ 23. $\dfrac{1 - \dfrac{1}{10^7}}{1 - \dfrac{1}{10}} = \frac{10}{9}\left(1 - \dfrac{1}{10^7}\right)$

25. $\dfrac{1 - (\frac{2}{3})^{11}}{1 - \frac{2}{3}} = 3[1 - (\frac{2}{3})^{11}]$

Section 0.3

1. (a) $>$ (b) $>$ (c) $>$ (d) $<$ (e) $<$
   (f) $<$ (g) $>$ (h) $>$ (i) $=$ (j) $=$
   (k) $>$

3. (b), (c), and (e)

5. (a)   T, because $15 = 15$ (b)   T, because $2 < 4$
    (c)   T   (d)   T   (e)   F; take $a = -1$ and $b = 2$
    (f)   F; take $a = -1$ and $b = 1$ (g)   T, because $4 < 5$
    (h)   T   (i)   T   (j)   F; take $a = 1$ and $b = -2$

CHAPTER 1

Section 1.1
1. $\{1, 2, 3, 4, 5\}$ 3. $\emptyset$ 5. A 7. $\emptyset$
9. $\{3, 4\} \cup \{5\} = \{3, 4, 5\}$

11. 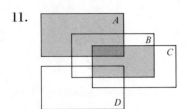   13.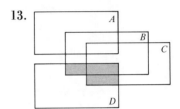

15. 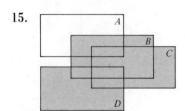   17.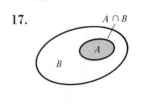

19. 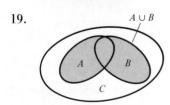   21.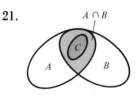

23.   Both statements say that A and B contain the same elements.
25.   Only $A = \emptyset$ is false; the others are true.
27. $a = b$
29. $\emptyset$ $\{0\}$ $\{1\}$ $\{2\}$ $\{0, 1\}$ $\{0, 2\}$ $\{1, 2\}$ $\{0, 1, 2\}$
31.   (a) $\{HHT, HTH, THH, HHH\}$ (b) $\{HTT, THT, TTH\}$
      (c) $\{HHH\}$
33.   (a) $\{$White-red, white-black, red-black$\}$
      (b) $\{$White-white, red-red, black-black, white-red,
            white-black, red-black, red-white, black-white, black-red$\}$
35. $S = \{\{a_1, b_1, c_1\}, \{a_1, b_1, c_2\}, \{a_1, b_2, c_1\}, \{a_1, b_2, c_2\},$
      $\{a_2, b_1, c_1\}, \{a_2, b_1, c_2\}, \{a_2, b_2, c_1\}, \{a_2, b_2, c_2\}\}$

Section 1.2

1. $\{1, 2\}$ 3. $\{1, 2, 3, 4, 6\}$ 5. A 7. $\emptyset$ 9. $\{5\}$
11. $\{6\}$

13. 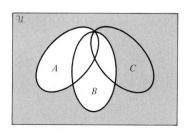   15. Same as Exercise 13.

17. 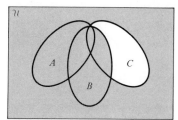   19.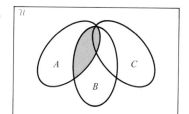

21. 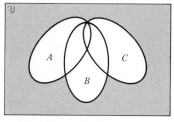   23.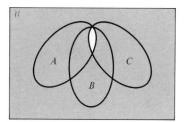

25.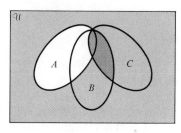

27. $540 + 1{,}768 = 2{,}308$ 29. $10{,}760 + 948 = 11{,}708$
31. 948 33. $29{,}324$
35. $\emptyset, \{a\}, \{b\}, \{c\}, \{a, b\}, \{a, c\}, \{b, c\}, \{a, b, c\}$
37. $\{1, 4\}$ 39. $\{0, 2, 3, 5, 6\}$ 41. A 43. $\emptyset$
45. $B - A$ 47. $B - (A \cup C)$ or $(B - A) \cap (B - C)$
49. $\overline{A \cup B \cup C}$

Section 1.3

3. $\{1, 2, 3, 4\}$ 5. $\emptyset$ 7. $\{2, 3\}$ 9. $\{4, 8, 12, 16, \ldots\}$

11. $\{1\}$ 13. $\emptyset$ 15. $\{5, 6, 7, 8, 9\}$ 17. $\mathcal{U}$

21. $\{x | x = 10 \text{ or } x > 10\}$ or $\{x | x < 10\}$

23. $\{x | x = 2^n \text{ where } n = 1, 2, 3, \ldots\}$

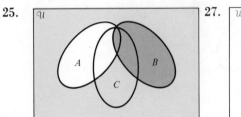

25.   27.

29.   31.

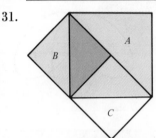

33.   35.

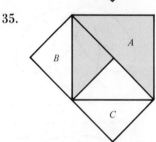

Section 1.4

1. F   3. F   5. T   7. $p(x)$ and $q(x)$ are unrelated

9. $q(x) \rightarrow p(x)$ 11. $q(x) \rightarrow p(x)$

13. $\{x | x^2 = x\} = \{x | x^2 - x = 0\} = \{x | x(x - 1) = 0\} = \{0, 1\}$

15. $\{x | x < -3\} \subset \{x | x < -2\}$ 17. $\{x | x > 2\} \subset \{x | x > 1\}$

19. $\{1962, 1967\}$ 21. $\{1962\}$ 23. $\{\text{Philadelphia, Cincinnati}\}$

25. $(A - B) \cup (B - A)$ 27. A 29. $A - B$

31. $A \cap \overline{B} = \overline{A} \cup B$ 33. (b) and (c) are correct

CHAPTER 2

Section 2.1

1. New Mexico, Tennessee

3. Florida, Louisiana, Pennsylvania   5. Florida, Louisiana

7. (1) $P \cap R$ (2) $Q \cup R$ (3) $\overline{Q}$ (4) $\overline{P \cup Q \cup R}$
   (5) $\overline{P \cap Q} = \overline{P} \cup \overline{Q}$ (6) $Q - (P \cup R)$
9. Nonsupport is and mental cruelty is not grounds for divorce in x.
11. Pregnancy at marriage and nonsupport are not grounds for divorce in x.
13. R is the set of states in which mental cruelty is grounds for divorce.
15. $\overline{P}$ is the set of states in which pregnancy at marriage is not grounds for divorce.
17. $Q - R$ is the set of states in which nonsupport but not mental cruelty is grounds for divorce.
19. (a) $\{2, 4, 6\}$ (b) $\{1, 3, 5\}$ (c) $\{1, 2, 3\}$ (d) $\emptyset$
    (e) $\{2\}$ (f) $\{1, 2, 3, 5\}$

Section 2.2

1. (a) The exercise is long and easy.
   (b) The exercise is not long.
   (c) The exercise is not long and it is easy.
   (d) The exercise is not long and it is not easy.
   (e) The exercise is not both long and easy.
   (f) The exercise is not both long and hard.
3. (a) $p \wedge q$ (b) $\sim(p \vee q)$ (c) $\sim p \wedge \sim q$
   (d) $p \wedge \sim q$ (e) $(q \vee \sim q) \wedge p$
5. p: "An integer is zero"; q: "An integer is even"; r: "An integer is odd"; $p \vee q \vee r$
7. p: "The answer to question 1 was yes"; q: "The answer to question 2 was no"; $p \wedge q$
9. $\{$Mercury, Mars$\}$ 11. $\emptyset$ 13. $\{$Uranus, Neptune, Pluto$\}$
15. $P \cap Q \cap \overline{R}$ 17. $P \cap (\overline{Q} \cup R) = P \cap \overline{(Q \cap R)}$
19. $P \cap (Q \cup R)$ 21. $\overline{P \cup Q} = \overline{P} \cap \overline{Q}$
23. $\overline{P \cup (Q \cap R)} = \overline{P} \cap \overline{(Q \cap R)} = \overline{P} \cap (\overline{Q} \cup \overline{R})$
25. (a) The outcome was two or three heads.
    (b) The outcome was not three heads.
    (c) The outcome was not two heads and not three heads.
    (d) The outcome was neighter two nor three heads.

Section 2.3

1.

p	q	p	$\wedge$	$\sim q$
T	T	T	F	F
T	F	T	T	T
F	T	F	F	F
F	F	F	F	T

3.

p	q	$\sim$	$(p$	$\vee$	$q)$
T	T	F	T	T	T
T	F	F	T	T	F
F	T	F	F	T	T
F	F	T	F	F	F

5.

p	q	$\sim$	$(p$	$\vee$	$q)$	$\wedge$	$(\sim p$	$\vee$	$q)$
T	T	F	T	T	T	F	F	T	T
T	F	F	T	T	F	F	F	F	F
F	T	F	F	T	T	F	T	T	T
F	F	T	F	F	F	T	T	T	F

7.

p	q	r	$(p \vee q) \wedge$		$\sim(p \wedge r)$	
T	T	T	T	F	F	T
T	T	F	T	T	T	F
T	F	T	T	F	F	T
T	F	F	T	T	T	F
F	T	T	T	T	T	F
F	T	F	T	T	T	F
F	F	T	F	F	T	F
F	F	F	F	F	T	F

9.

p	q	r	$(p \wedge q) \vee$		$(p \wedge r)$
T	T	T	T	T	T
T	T	F	T	T	F
T	F	T	F	T	T
T	F	F	F	F	F
F	T	T	F	F	F
F	T	F	F	F	F
F	F	T	F	F	F
F	F	F	F	F	F

11.

p	q	$\sim(\sim p$	$\vee$	$\sim q)$	
T	T	T	F	F	F
T	F	F	F	T	T
F	T	F	T	T	F
F	F	F	T	T	T

13.   1 point   15.   false

17.

p	q	r	t
T	T	T	T
T	T	T	F
T	T	F	T
T	T	F	F
T	F	T	T
T	F	T	F
T	F	F	T
T	F	F	F
F	T	T	T
F	T	T	F
F	T	F	T
F	T	F	F
F	F	T	T
F	F	T	F
F	F	F	T
F	F	F	F

19.

p	q	$\sim(p \wedge q)$		$\sim p \vee \sim q$	
T	T	F	T	F F	F
T	F	T	F	F T	T
F	T	T	F	T T	F
F	F	T	F	T T	T

21.

p	$\sim(\sim p)$	
T	T	F
F	F	T

23.

p	q	$\sim(p \wedge \sim q)$		$\sim p \vee q$	
T	T	T T F F		F	T T
T	F	F T T T		F	F F
F	T	T F F F		T	T T
F	F	T F F T		T	T F

25.

p	q	r	$(p \lor q) \land r$			$(p \land r) \lor (q \land r)$		
T	T	T	T	T T		T	T	T
T	T	F	T	F F		F	F	F
T	F	T	T	T T		T	T	F
T	F	F	T	F F		F	F	F
F	T	T	T	T T		F	T	T
F	T	F	T	F F		F	F	F
F	F	T	F	F T		F	F	F
F	F	F	F	F F		F	F	F

Section 2.4

1.

p	q	$\sim p \to q$		
T	T	F	T T	
T	F	F	T F	
F	T	T	T T	
F	F	T	F F	

3.

p	q	$\sim p \to \sim q$		
T	T	F	T	F
T	F	F	T	T
F	T	T	F	F
F	F	T	T	T

5.

p	q	$(p \land q) \to q$		
T	T	T	T T	
T	F	F	T F	
F	T	F	T T	
F	F	F	T F	

7.

p	q	$(\sim p \land \sim q) \Leftrightarrow (p \lor q)$				
T	T	F	F	F	F	T
T	F	F	F	T	F	T
F	T	T	F	F	F	T
F	F	T	T	T	F	F

9.

p	q	r	$(p \Leftrightarrow q) \rightarrow (p \Leftrightarrow r)$		
T	T	T	T	T	T
T	T	F	T	F	F
T	F	T	F	T	T
T	F	F	F	T	F
F	T	T	F	T	F
F	T	F	F	T	T
F	F	T	T	F	F
F	F	F	T	T	T

11.

p	q	r	$(p \rightarrow q) \rightarrow (p \rightarrow r)$		
T	T	T	T	T	T
T	T	F	T	F	F
T	F	T	F	T	T
T	F	F	F	T	F
F	T	T	T	T	T
F	T	F	T	T	T
F	F	T	T	T	T
F	F	F	T	T	T

13.

p	q	$p \rightarrow q$	$\sim q \rightarrow \sim p$		
T	T	T	F	T	F
T	F	F	T	F	F
F	T	T	F	T	T
F	F	T	T	T	T

15. $\sim p \rightarrow \sim q$, $\sim q \vee p$, and $\sim(\sim p \wedge q)$

17. The statement is true in the indicated cases, where a stands for "switch A is closed", b for "switch B is closed", and c for "both lights are on".

a	b	c	$c \Leftrightarrow (a \wedge b)$		
→T	T	T	T T	T
T	T	F	F F	T
T	F	T	T F	F
→T	F	F	F T	F
F	T	T	T F	F
→F	T	F	F T	F
F	F	T	T F	F
→F	F	F	F T	F

19.

p	q	$p \rightarrow q$	$\sim p \vee q$	
T	T	T	F	T T
T	F	F	F	F F
F	T	T	T	T T
F	F	T	T	T F

21.

p	q	$\sim(p \rightarrow q)$		$p \wedge \sim q$		
T	T	F	T	T F	F
T	F	T	F	T T	T
F	T	F	T	F F	F
F	F	F	T	F F	T

23.

p	q	r	$(p \rightarrow q) \wedge r$			$(\sim p \vee q) \wedge r$		
T	T	T	T	T T		T	T T	
T	T	F	T	F F		T	F F	
T	F	T	F	F T		F	F T	
T	F	F	F	F F		F	F F	
F	T	T	T	T T		T	T T	
F	T	F	T	F F		T	F F	
F	F	T	T	T T		T	T T	
F	F	F	T	F F		T	F F	

25.

p	q	r	$\sim[(p \rightarrow q) \wedge r]$				$(p \wedge \sim q) \vee \sim r$		
T	T	T	F	T	T T		F	F	F
T	T	F	T	T	F F		F	T	T
T	F	T	T	F	F T		T	T	F
T	F	F	T	F	F F		T	T	T
F	T	T	F	T	T T		F	F	F
F	T	F	T	T	F F		F	T	T
F	F	T	F	T	T T		F	F	F
F	F	F	T	T	F F		F	T	T

Section 2.5

3. $p \wedge (q \vee r)$ 5. $(p \wedge \sim q) \vee (\sim q \wedge p) = p \wedge \sim q$

7. $(p \wedge r) \vee (q \wedge r) = (p \vee q) \wedge r$ 11. $P' + (P + R)$

13. $(P + Q') \cdot (P' + Q)$

15. This statement is equivalent to $(p \vee q) \wedge \sim(p \vee q)$. To see this, first write $\sim(p \vee q) \Leftrightarrow (p \vee q)$, then $[\sim(p \vee q) \rightarrow (p \vee q)] \wedge [(p \vee q) \rightarrow \sim(p \vee q)]$, then $[(p \vee q) \vee (p \vee q)] \wedge [\sim(p \vee q) \vee \sim(p \vee q)]$ and this gives the above statement. Hence the switching network can be taken to be $(P + Q) \cdot (P + Q)'$.

17. $P \cdot Q \cdot R' + P \cdot Q' \cdot R + P' \cdot Q \cdot R + P' \cdot Q' \cdot R'$

19.

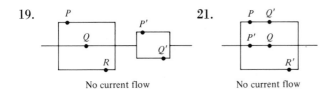

No current flow

21.

No current flow

23.

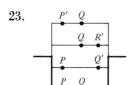

Current flows as indicated.

25.

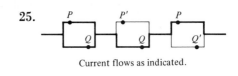

Current flows as indicated.

Section 2.6

1. $p \to (a \lor b)$
   a
   $\therefore \sim b$

   This argument is invalid (let a, b, and p be True); a stands for "$A = 0$", b for "$B = 0$", and c for "$AB = 0$".

3. $a \lor b$
   $a \to c$
   $c \to b$
   $\therefore a \land b$

   This argument is invalid (let a and c be False, b be True); a stands for "set A is empty", b for "set B is empty", and c for "set C is empty".

5. $a \Leftrightarrow b$
   $\sim b$
   $\therefore \sim a$

   This argument is valid; a stands for "the light is on", b for "the switch is on".

7. $s \Leftrightarrow d$
   $\sim r \to \sim d$
   r
   $\therefore s$

   This argument is invalid (let s and d be False, r be True); s stands for "seawater can be used for drinking", d for "it is desalted", and r for "appropriate research is carried out".

9. The argument is valid.

p	q	r	$p \Leftrightarrow q$	$q \Leftrightarrow r$	$p \Leftrightarrow r$
T	T	T	T	T	T
T	T	F	T	F	F
T	F	T	F	F	T
T	F	F	F	T	F
F	T	T	F	T	F
F	T	F	F	F	T
F	F	T	T	F	F
F	F	F	T	T	T

11. The argument is invalid.

p	q	$p \to \sim q$	$q \to p$	$p \wedge q$
T	T	F	T	T
T	F	T	T	F←
F	T	T	F	F
F	F	T	T	F←

13. $[(p \vee q) \wedge \sim p] \to q$ is equivalent to $\sim[(p \vee q) \wedge \sim p] \vee q$, or
$\sim(p \vee q) \vee p \vee q$, or $(\sim p \wedge \sim q) \vee p \vee q$. The switching network is $P' \cdot Q' + P + Q$.

15. We have $\sim[(p \vee q) \wedge (\sim q \vee r)] \vee (\sim p \vee r)$, or
$(p \wedge \sim q) \vee (q \wedge \sim r) \vee (\sim p \vee r)$. The switching network is
$P \cdot Q' + Q \cdot R' + P' + R$.

CHAPTER 3

Section 3.1

1. $n = 45 + 72 - 28 = 89$

3. $8 + 7 + 8 + 4 + 7 = 34$ 5.   164

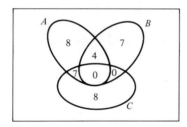
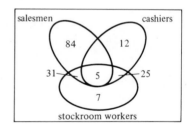

7. (a) $n(A) + n(B) - n(A \cap B) = 15 + 10 - 5 = 20$
(b) $10 + 6 - 2 = 14$
(c) $n(A) + n(B) + n(C) - n(A \cap B) - n(A \cap C)$
$- n(B \cap C) + n(A \cap B \cap C)$
$= 15 + 10 + 6 - 5 - 3 - 2 + 1 = 22$

9. (a) The sets are mutually disjoint.
(b) The sets have the same elements; that is, they are equal.
(c) $C \subset A \cup B$

11. Each question can be answered in three ways independently of
the other questions. There are therefore 3^{10} different answers
(of which only one is correct).

Section 3.2

1. (a) F   (b) T   (c) T   (d) F   (e) T
(f) T   (g) F   (h) F

3. $2 \times 9! = 725,760$

5. (a) one way   (b) This cannot happen, because at least two notices have to be in wrong envelopes if not all are in the right envelopes.

   (c) $\dfrac{250 \times 249}{2} = 31,125$

7. $2 \times 7 \times 3 = 42$ 9. $\dfrac{6 \times 5}{2} \times 8 \times 7 \times 6 \times 5 = 25,200$

11. $5 \times 4 \times 7! + 4 \times 3 \times 7! = 32 \times 7! = 161,280$

13. $28!$ 15. $27!$ 17. $8! = 40,320$

19.

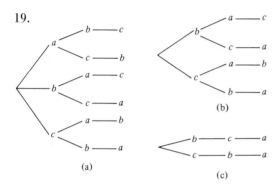

(a)

(b)

(c)

Section 3.3

1. $9!$ 3. (a) one way   (b) $4! = 24$ 5. $\dfrac{7!}{(2!)^3} = 630$

7. $\dfrac{10!}{3! \, 3! \, 2!} = 50,400$ 9. $9 \times 10 = 90$ 11. $\dfrac{98!}{(14!)^7}$

Section 3.4

1. (a) $\dbinom{5}{0} = \dbinom{5}{5} = 1$ $\dbinom{5}{1} = \dbinom{5}{4} = 5$ $\dbinom{5}{2} = \dbinom{5}{3} = 10$

   (b) $\dbinom{6}{0} = \dbinom{6}{6} = 1$ $\dbinom{6}{1} = \dbinom{6}{5} = 6$ $\dbinom{6}{2} = \dbinom{6}{4} = 15$

   $\dbinom{6}{3} = 20$

   (c) 101 (d) $\dfrac{100 \times 99 \times 98}{6} = 161,700$ (e) 0

   (f) 1 (g) 1

3. $\dbinom{10,000}{5} = \dfrac{10,000!}{5! \, 9,995!}$ 5. $\dbinom{70}{10}\dbinom{60}{10} = \dfrac{70!}{10! \, 10! \, 50!}$

7. $\dbinom{20}{8}$ 9. $4\dbinom{13}{5} = 5,148$

11. $\binom{13}{1}\binom{13}{1}\binom{13}{1}\binom{13}{10} = 13^3 \times 286 = 628{,}342$

13. $\binom{39}{13}$ 15. $\dfrac{n(n-3)}{2}$ 17. $\binom{5}{2}\binom{21}{3} = 13{,}300$

19. $\binom{9{,}920}{90}\binom{80}{10} + \binom{9{,}920}{89}\binom{80}{11} + \cdots + \binom{9{,}920}{20}\binom{80}{80}$

Section 3.5

1. $x^4 + 4x^3y + 6x^2y^2 + 4xy^3 + y^4$
3. $x^6 + 6x^5y + 15x^4y^2 + 20x^3y^3 + 15x^2y^4 + 6xy^5 + y^6$
5. $x^5 - 5x^4y + 10x^3y^2 - 10x^2y^3 + 5xy^4 - y^5$
7. $64x^6 + 192x^5y + 240x^4y^2 + 160x^3y^3 + 60x^2y^4 + 12xy^5 + y^6$
9. $x^5 + 5x^3 + 10x + 10\dfrac{1}{x} + 5\dfrac{1}{x^3} + \dfrac{1}{x^5}$
11. $\frac{625}{256} \sim 2.4414$ 13. 1.04060401
15. (a) $\binom{40}{3} = 9{,}880$ (b) $\dfrac{40!}{20!\,20!}$ (c) $\dfrac{40!}{10!\,30!}$
17. $\binom{10}{0} + \binom{10}{1} + \binom{10}{2} + \binom{10}{3} = 1 + 10 + 45 + 120 = 176$
19. The left side equals the binomial expansion of $(1-1)^n$.
21. $2^{20} = 1{,}024^2 = 1{,}048{,}576$

Section 3.6

1. (a) and (d)
3. (a) 380   (b) 20   (c) 1,140   (d) 105   (e) 1
   (f) $\dfrac{40 \times 39 \times 38 \times 37}{2 \times 2} = 548{,}340$
5. $\dfrac{1}{4!} \times \dfrac{48!}{9!(13!)^3}$ partitions; $\dfrac{48!}{9!(13!)^3}$ ordered partitions
7. $\dfrac{52!}{47!}$
9. [{Michigan, New Mexico, Tennessee, Wisconsin}, {Florida, Louisiana, Massachusetts, Pennsylvania, Vermont}]
11. (a) [{TTT, TTF, TFT, FTT, FFT}, {TFF, FTF, FFF}]
    (b) [{TTT, TTF, TFT}, {TFF, FTT, FTF, FFT, FFF}]
    (c) [{TTF, TFF, FTT, FTF}, {TTT, TFT, FFT, FFF}]
    (d) [{TTT, TFT, TTF, FTF, FFF}, {FTT, FFT, TFF}]
    (e) [{TTF, FTT, FFT, FFF}, {TTT, TFT, TFF, FTF}]
    (f) [{TTT, TFT, TFF, FTT, FFT, FFF}, {TTF, FTF}]
    (g) [{TTT, TTF, TFT, TFF, FTT, FFF}, {FTF, FFT}]

Section 3.7

1. three times
3. Number the states 1, 2, ..., 9. Questions: Is it 1, 2, 3, or 4? If yes, partition further; if no, is it 5, 6, 7, or 8? If not, it was state 9;

if yes, partition further. In any case, you need no more than four questions.

5. 4

9. The following diagram indicates a sequence of tests where you move to the next step if the lights do *not* go on. This gives you a group of four bulbs with the two defective bulbs which you proceed to partition. Whenever the lights go on, you also located a group of four bulbs with the defective ones. You may need three more tests to locate the two defective bulbs.

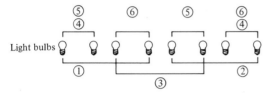

CHAPTER 4

Section 4.1

1. (a) $S = \{HH, HT, TH, TT\}$ (b) $\frac{1}{4}$ (c) $\frac{2}{4} = \frac{1}{2}$
   (d) $\frac{3}{4}$

3. Yes, because the respective probabilities are $\frac{1}{16}$ and $\frac{1}{32}$.

5. (a), (b), and (e)

7. (a) $S = \{2, 3, 4, \ldots, 12\}$
   (b) Let $w(k)$ be the weight of the outcome when k is the sum of the showing dots. Then $w(2) = w(12) = \frac{1}{36}$, $w(3) = w(11) = \frac{2}{36}$, $w(4) = w(10) = \frac{3}{36}$, $w(5) = w(9) = \frac{4}{36}$, $w(6) = w(8) = \frac{5}{36}$, and $w(7) = \frac{6}{36}$.

9. $\dfrac{1}{2,000}$

11. (a) $\frac{2}{7}$ (b) $\frac{5}{7}$ (c) 1

13. (a) $\frac{1}{2}$ (b) $\frac{1}{2}$

15. (a) $w_1 + w_5 = \frac{17}{31}$
    (b) He should pick question q_5 because then his probability for winning the prize is $\frac{16}{31} > \frac{15}{31}$.

17. (a) $13 \times \dfrac{3}{4 \times 26} = \dfrac{3}{8}$ (b) $2\left(\dfrac{3}{4 \times 26} + \dfrac{1}{4 \times 26}\right) = \dfrac{1}{13}$

   (c) $24\left(\dfrac{3}{4 \times 26} + \dfrac{1}{4 \times 26}\right) = \dfrac{12}{13}$

   (d) $13\left(\dfrac{3}{4 \times 26} + \dfrac{1}{4 \times 26}\right) = \dfrac{1}{2}$ (e) $\frac{1}{4} + 2 \times \dfrac{3}{4 \times 26} = \dfrac{4}{13}$

Section 4.2

1. (a) $\frac{2}{10} = \frac{1}{5}$ (b) $\frac{7}{10}$ (c) 1

3. (a) $\frac{1}{10}$ (b) $\frac{1}{5}$

5. (a) $\dfrac{12 + 17 - 3}{150} = \dfrac{13}{75}$ (b) $\dfrac{14}{150} = \dfrac{7}{75}$ (c) $1 - \dfrac{13}{75} = \dfrac{62}{75}$

7. (a) $1 - \dfrac{52 + 27 - 19}{100} = \dfrac{2}{5}$ (b) $\dfrac{19}{100}$

   (c) $\dfrac{52 + 27 - 19}{100} = \dfrac{3}{5}$ (d) $\dfrac{52 - 19}{100} = \dfrac{33}{100}$

9. (a) $\dfrac{3}{36} = \dfrac{1}{12}$ (b) $\dfrac{9}{36} = \dfrac{1}{4}$ (c) $\dfrac{6}{36} + \dfrac{2}{36} = \dfrac{2}{9}$

11. (a) $\dfrac{10}{24} = \dfrac{5}{12}$ (b) $\dfrac{18}{24} = \dfrac{3}{4}$ (c) $\dfrac{5}{24}$ (d) $\dfrac{1}{24}$

Section 4.3

1. (a) $\frac{4}{9}$ (b) $\frac{5}{8}$ (c) no

3. (a) Most likely is c_4; least likely are c_1 and c_7.
   (b) $\frac{1}{9}$ (c) c_3

5. $\dfrac{6}{16 + 6} = \dfrac{3}{11}$

7. (a) $\dfrac{1}{2^{10}}$ (b) $\dfrac{1}{2} \times \dfrac{1}{2^6} = \dfrac{1}{2^7}$

9. (a) dependent   (b) dependent   (c) independent, be-
   cause since $P(A \cap B) = \frac{1}{8}$, $P(A) = \frac{2}{8}$, $P(B) = \frac{4}{8}$, and so
   $\frac{1}{8} = \frac{2}{8} \times \frac{4}{8}$

Section 4.4

1. $P(A_2|B) = \dfrac{\frac{17}{26} \times \frac{3}{100}}{\frac{9}{26} \times \frac{1}{100} + \frac{17}{26} \times \frac{3}{100}} = \dfrac{51}{60} = 0.85$

3. (a) $P(A_2|B) = \frac{4}{17}/\frac{8}{17} = \frac{1}{2}$
   (b) $P(A_1 \cup A_2) = \frac{13}{17}$, $P(B|A_1 \cup A_2) = \frac{8}{13}$; hence
   $$P(A_1 \cup A_2|B) = \dfrac{\frac{13}{17} \times \frac{8}{13}}{\frac{13}{17} \times \frac{8}{13} + 0} = 1$$

5. $P(A_1) = \frac{6}{17}$ $P(B|A_1) = \frac{2}{6}$ $P(C|A_1) = \frac{1}{6}$ $P(D|A_1) = \frac{2}{6}$
   $P(A_2) = \frac{3}{17}$ $P(B|A_2) = 0$ $P(C|A_2) = \frac{1}{3}$ $P(D|A_2) = 1$
   $P(A_3) = \frac{8}{17}$ $P(B|A_3) = 0$ $P(C|A_3) = \frac{5}{8}$ $P(D|A_3) = 1$
   (a) $P(A_1|B) = 1$

   (b) $P(A_1|C) = \dfrac{\frac{6}{17} \times \frac{1}{6}}{\frac{7}{17}} = \frac{1}{7} \approx 0.143$

   (c) $P(A_1 \cup A_2|B) = \dfrac{\frac{9}{17} \times \frac{2}{9}}{\frac{9}{17} \times \frac{2}{9} + \frac{8}{17} \times 0} = 1$

   (d) $P(A_2|C) = \frac{1}{7} \approx 0.143$
   (e) $P(A_2|D) = \frac{3}{13} \approx 0.231$
   (f) $P(A_3|C) = \frac{5}{7} \approx 0.714$
   (g) $P(A_3|D) = \frac{8}{13} \approx 0.615$

   (h) $P(A_2 \cup A_3|C) = \dfrac{\frac{11}{17} \times \frac{6}{11}}{\frac{11}{17} \times \frac{6}{11} + \frac{6}{17} \times \frac{2}{6}} = \frac{6}{8} = 0.75$

   (i) $P(A_2 \cup A_3|D) = \dfrac{\frac{11}{17} \times 1}{\frac{11}{17} \times 1 + \frac{6}{17} \times \frac{2}{6}} = \frac{11}{13} \approx 0.846$

7. Let the output of machines a, b, and c be A, B, and C, respectively; let D stand for the defective sample. Then

$$P(A) = \tfrac{40}{100} \qquad P(D|A) = \tfrac{5}{40}$$
$$P(B) = \tfrac{25}{100} \qquad P(D|B) = \tfrac{2}{25}$$
$$P(C) = \tfrac{35}{100} \qquad P(D|C) = \tfrac{1}{35}$$
$$P(A)P(D|A) + P(B)P(D|B) + P(C)P(D|C) = \tfrac{8}{100}$$

(a) $\quad P(A|D) = \dfrac{\tfrac{40}{100} \times \tfrac{5}{40}}{\tfrac{8}{100}} = \tfrac{5}{8}$

(b) $\quad P(C|D) = \dfrac{\tfrac{35}{100} \times \tfrac{1}{35}}{\tfrac{8}{100}} = \tfrac{1}{8}$

(c) $\quad P(A \cup B|D) = \dfrac{\tfrac{65}{100} \times \tfrac{7}{65}}{\tfrac{65}{100} \times \tfrac{7}{65} + \tfrac{35}{100} \times \tfrac{1}{35}} = \tfrac{7}{8}$

9. $P(R|F) = \tfrac{93}{100} \qquad P(R|M) = \tfrac{88}{100}$

$$P(F)P(R|F) + P(M)P(R|M) = \tfrac{60}{100} \times \tfrac{93}{100} + \tfrac{40}{100} \times \tfrac{88}{100} = \tfrac{91}{100}$$

$$P(F|R) = \dfrac{\tfrac{60}{100} \times \tfrac{93}{100}}{\tfrac{91}{100}} \approx 0.613$$

Section 4.5

1. (a) $\dbinom{7}{7}\left(\dfrac{2}{3}\right)^7\left(\dfrac{1}{3}\right)^0 = \left(\dfrac{2}{3}\right)^7 \approx 0.0585$

   (b) $\dbinom{7}{6}\left(\dfrac{1}{3}\right)^6\left(\dfrac{2}{3}\right) = \dfrac{14}{3^7} \approx 0.0064$

   (c) $\dfrac{14}{3^7} + \dfrac{1}{3^7} = \dfrac{15}{3^7} \approx 0.0068$

3. $\dbinom{5}{1}\left(\dfrac{5}{100}\right)\left(\dfrac{95}{100}\right)^4 = \dfrac{1}{4}\left(\dfrac{19}{20}\right)^4 \approx 0.20$

5. (a) those with 5 or 6 children

   (b) those with 10 children. This is seen from the accompanying table.

n	$\dbinom{n}{3}\dfrac{1}{2^n}$
3	$\tfrac{1}{8} = 0.125$
4	$\tfrac{1}{4} = 0.25$
5	$\tfrac{5}{16} = 0.3125$
6	$\tfrac{5}{16} = 0.3125$
7	$\tfrac{35}{128} \approx 0.273$
8	$\tfrac{7}{32} \approx 0.219$
9	$\tfrac{21}{128} \approx 0.164$
10	$\tfrac{15}{128} \approx 0.117$

7. (a) $\dbinom{5}{5}\left(\dfrac{1}{6}\right)^5\left(\dfrac{5}{6}\right)^0 = \dfrac{1}{6^5} \approx 0.00013$

   (b) $\dbinom{5}{4}\left(\dfrac{1}{6}\right)^4\left(\dfrac{5}{6}\right) = \dfrac{25}{6^5} \approx 0.0032$

(c) $\binom{5}{1}\left(\frac{1}{6}\right)\left(\frac{5}{6}\right)^4 = \left(\frac{5}{6}\right)^5 \approx 0.40$

9.

m	$\dfrac{m!}{\left(\dfrac{m}{2}\right)!\left(\dfrac{m}{2}\right)!}(\frac{1}{2})^m$
4	$\frac{3}{8} = 0.375$
8	$\frac{35}{128} \approx 0.273$
12	$\frac{231}{1024} \approx 0.226$
16	$\dfrac{16!}{8!\,8!} \cdot \dfrac{1}{2^{16}} \approx 0.196$
20	$\dfrac{20!}{10!\,10!} \cdot \dfrac{1}{2^{20}} \approx 0.176$

11. (a) $\binom{5}{5}\left(\frac{1}{28}\right)^5\left(\frac{27}{28}\right)^0 = \frac{1}{28^5}$

    (b) $\binom{5}{3}\left(\frac{1}{28}\right)^3\left(\frac{27}{28}\right)^2 = \frac{10 \times 27^2}{28^5} \approx 0.0004$

    (c) $\binom{5}{5}\left(\frac{7}{28}\right)^5\left(\frac{21}{28}\right)^0 = \left(\frac{1}{4}\right)^5 \approx 0.001$

    (d) $\binom{5}{0}\left(\frac{7}{28}\right)^0\left(\frac{21}{28}\right)^5 = \left(\frac{3}{4}\right)^5 \approx 0.237$

    (e) $\binom{5}{0}\left(\frac{1}{28}\right)^0\left(\frac{27}{28}\right)^5 = \left(\frac{27}{28}\right)^5 \approx 0.865$

    (f) $\binom{5}{2}\left(\frac{3}{28}\right)^2\left(\frac{25}{28}\right)^3 = \frac{10 \times 9 \times 625}{28^5} \approx 0.0033$

13. (a) $\binom{6}{6}\left(\frac{4}{5}\right)^6\left(\frac{1}{5}\right)^0 = \left(\frac{4}{5}\right)^6 \approx 0.262$

    (b) $\binom{6}{6}\left(\frac{1}{5}\right)^6\left(\frac{4}{5}\right)^0 = \frac{1}{5^6} \approx 0.000064$

    (c) $\binom{6}{2}\left(\frac{1}{5}\right)^2\left(\frac{4}{5}\right)^4 = \frac{15 \times 256}{5^6} \approx 0.246$

Section 4.6

1. (a) $\frac{1}{3} \times \frac{1}{3} \times \frac{1}{2} = \frac{1}{18}$ (b) $\frac{1}{3} \times \frac{1}{2} \times \frac{3}{4} = \frac{1}{8}$

  (c) $\frac{1}{3} \times \frac{1}{2} \times 1 = \frac{1}{6}$ (d) $1 - \frac{1}{3} \times \frac{1}{3} \times \frac{1}{3} = \frac{26}{27}$

  (e) $1 - \frac{1}{3} \times \frac{1}{3} \times 1 = \frac{8}{9}$

3. (a) $\frac{1}{4}\left(\frac{5}{1,000} + \frac{8}{1,000}\right) = \frac{13}{4,000}$

  (b) $\binom{400,000}{1,800}\left(\frac{18}{1,000}\right)^{1,800}\left(\frac{982}{1,000}\right)^{398,200}$

(c) $\binom{400,000}{900}\left(\frac{18}{1,000}\right)^{900}\left(\frac{982}{1,000}\right)^{399,100}$

5. (a)

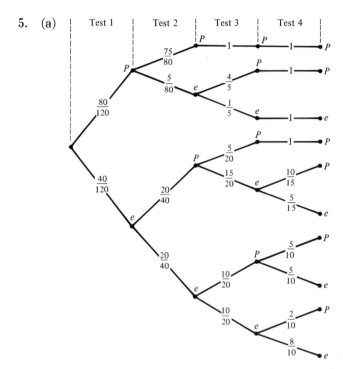

	Test 1	Test 2	Test 3	Test 4

(b) $\frac{80}{120} \times \frac{75}{80} \times 1 \times 1 = \frac{5}{8}$

(c) $\frac{40}{120} \times \frac{20}{40} \times \frac{10}{20} \times \frac{8}{10} = \frac{8}{120} = \frac{1}{15}$

(d) $\frac{20}{40} \times \frac{15}{20} + \frac{20}{40} \times \frac{7}{20} = \frac{11}{20}$

(e) $\frac{10}{15} + \frac{2}{10} = \frac{13}{15}$

Section 4.7

1. $E = 60 \times \frac{2}{10} + 70 \times \frac{5}{10} + 80 \times \frac{2}{10} + 90 \times \frac{1}{10} = 72$

3. The possible winnings are calculated as $(1 + 3 + 5) \times 4 = 36$ and the possible losses are $(2 + 4 + 6) \times 3 = 36$. Hence, $E = 36 \times \frac{1}{2} - 36 \times \frac{1}{2} = 0$, and so the game is fair.

5. (a) The total number of possible heads equals the total number of possible tails. Hence, $E = 0$.

    (b) $E = 0$ for the reason given in (a).

    (c) There are four possible outcomes with three Heads, and twelve other possible outcomes. Hence, we calculate
    $E = 3 \times \frac{4}{16} - 1 \times \frac{12}{16} = 0$.

    (d) $E = 4 \times \frac{4}{16} - 1 \times \frac{12}{16} = \frac{1}{4}$

CHAPTER 5

Section 5.1

1.

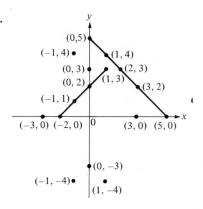

3.

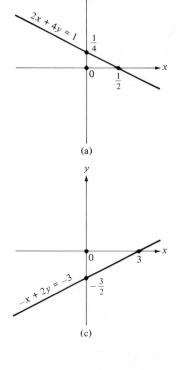

(a)

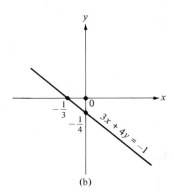

(b)

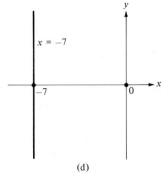

(c)

(d)

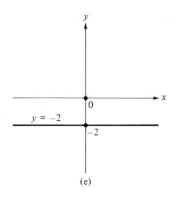

(e)

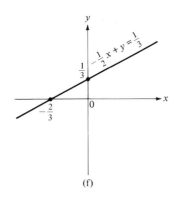

(f)

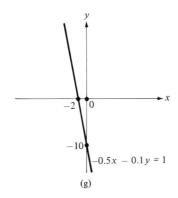

(g)

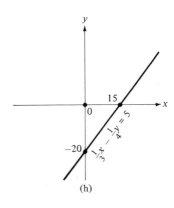

(h)

5. $(2, 3)$ and $(1, 4)$ 7. $(1, 2)$ or $(2, 3)$ or $(3, 4)$

Section 5.2

1. (a) $\dfrac{-2}{2} = -1$ (b) 2 (c) -1 (d) 0
   (e) undefined   (f) 0
3. $y + 2 = -x$, or $y = -x - 2$
5. $y - 5 = 2(x - 1)$, or $y = 2x + 3$
7. $y = \frac{3}{4}x - 3$
9. (a) no   (b) no   (c) yes
11. (a) $L(0) = -1$ $L(1) = 2$ $L(-1) = -4$
    (b) $M(\frac{5}{2}) = 0$ $M(1) = 3$ $M(-1) = 7$
    (c) $N(0) = 3$ $N(-2) = 3$ $N(3) = 3$
    (d) $Q(1) = 0$ $Q(-1) = -8$ $Q(-2) = -12$
    (e) $R(1) = -8$ $R(-1) = 0$ $R(\frac{1}{4}) = -5$
    (f) $S(-5) = -5$ $S(5) = 5$ $S(7) = 7$

Section 5.3

1. $(1, 0)$ 3. $(3, 1)$ 5. $(0, 0)$ 7. $\left(-\frac{20}{7}, \frac{39}{7}\right)$
9. $\left(\frac{6}{19}, \frac{3}{19}\right)$ 11. $(6, 3)$

13. (a) 450   (b) 405 dollars
    (c) $5,000 \times 0.4 - 180 \times 4 = 1,280$ dollars
15. (a) the second process   (b) the first process
17. $p = 63.64$ (rounded off); $q = 564$ (rounded off to the next unit)
19. $p = 5$; $q = 5.5$

Section 5.4

1.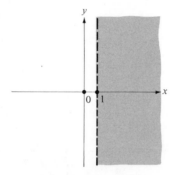

3.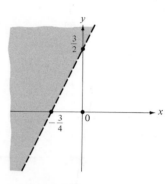

5.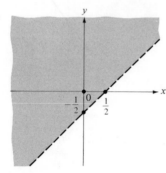

7.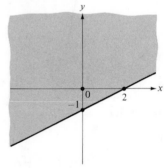

9.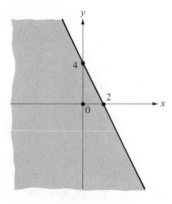

11.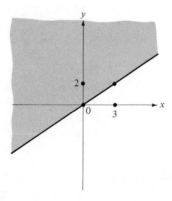

13.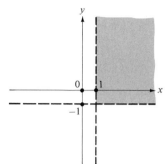

15.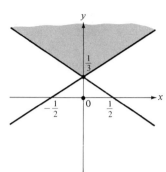

17.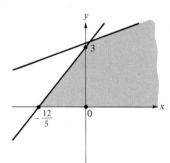

19.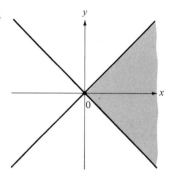

21.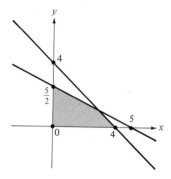

23.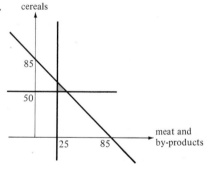

Section 5.5

1. (a) $P = (1250, 0)$ (b) $P = (750, 750)$
   (c) $P = (0, 1087.5)$ (d) $P = (0, 1087.5)$

3. (a) $P = (160, 0)$ (b) $P = (160, 0)$
   (c) $P = (\frac{480}{7}, \frac{960}{7})$ (d) $P = (0, 160)$

5. (a) $P = (4, 0)$ (b) $P = (2, 2)$
   (c) $P = (\frac{32}{5}, \frac{12}{5})$ (d) $P = (\frac{32}{5}, \frac{12}{5})$

7. With x pounds Cheddar and y pounds Cheddar spread produced, $f = [4 - (0.5 + 0.5 + 0.2)]x + [8 - (0.2 + 0.5 + 1)]y = 2.8x + 6.3y$ is to be maximized subject to the constraints

$$0.5x + 0.2y \leq 1,200$$
$$0.5x + 0.5y \leq 1,500$$
$$0.2x + \quad y \leq 2,200$$

The function is maximized at $P = (2000, 1000)$.

CHAPTER 6

Section 6.1

1. $[-2 \quad -2 \quad 4]$
3. $[\frac{1}{2} \quad -\frac{2}{3} \quad \frac{5}{6}]$

5. $\begin{bmatrix} 5 \\ 15 \\ 4 \end{bmatrix}$

7. $\begin{bmatrix} 1 & 1 \\ 1 & 1 \end{bmatrix}$

9. $\begin{bmatrix} 10 & -5 \\ 2 & 0 \\ 5 & 1 \end{bmatrix}$

11. $\begin{bmatrix} 7 & 5 & 0 & 0 \\ -3 & 2 & 0 & 0 \\ 0 & 0 & 6 & 0 \\ 0 & 0 & 0 & -2 \end{bmatrix}$

13. (a) $X = -\frac{1}{2}A = \begin{bmatrix} 1 & -\frac{1}{2} \\ -\frac{1}{2} & \frac{3}{2} \end{bmatrix}$

    (b) $X = E - D = \begin{bmatrix} \frac{1}{2} & -\frac{2}{3} \\ \frac{4}{3} & 0 \\ 0 & -1 \end{bmatrix}$

    (c) $X = 2C - F = \begin{bmatrix} -\frac{1}{3} & \frac{1}{3} & \frac{5}{3} \\ \frac{1}{2} & \frac{3}{2} & \frac{1}{2} \\ 1 & 0 & 0 \end{bmatrix}$

    (d) $X = -D = \begin{bmatrix} 0 & -1 \\ 1 & 0 \\ 0 & -1 \end{bmatrix}$

    (e) $X = F$ (f) $X = -A = \begin{bmatrix} 2 & -1 \\ -1 & 3 \end{bmatrix}$

Section 6.2

1. (a) $A \cdot B = \begin{bmatrix} 5 & 11 \\ 11 & 25 \end{bmatrix}$ (b) $B \cdot A = \begin{bmatrix} 6 & 14 \\ 14 & 20 \end{bmatrix}$

    (c) $A \cdot I = A$ (d) $I \cdot A = A$ (e) $A \cdot J = \begin{bmatrix} 2 & 1 \\ 4 & 3 \end{bmatrix}$

    (f) $J \cdot J = \begin{bmatrix} 1 & 0 \\ 0 & 1 \end{bmatrix}$ (g) $I \cdot I = I$

    (h) $(I + J) \cdot (I + J) = \begin{bmatrix} 1 & 1 \\ 1 & 1 \end{bmatrix} \cdot \begin{bmatrix} 1 & 1 \\ 1 & 1 \end{bmatrix} = \begin{bmatrix} 2 & 2 \\ 2 & 2 \end{bmatrix}$

    (i) $(A + I) \cdot J = \begin{bmatrix} 2 & 2 \\ 3 & 5 \end{bmatrix} \cdot \begin{bmatrix} 0 & 1 \\ 1 & 0 \end{bmatrix} = \begin{bmatrix} 2 & 2 \\ 5 & 3 \end{bmatrix}$

(j) $(2A + 3B) \cdot A = \begin{bmatrix} 5 & 13 \\ 12 & 20 \end{bmatrix} \cdot \begin{bmatrix} 1 & 2 \\ 3 & 4 \end{bmatrix} = \begin{bmatrix} 44 & 62 \\ 72 & 104 \end{bmatrix}$

(k) $(2I + 3J) \cdot J = \begin{bmatrix} 2 & 3 \\ 3 & 2 \end{bmatrix} \begin{bmatrix} 0 & 1 \\ 1 & 0 \end{bmatrix} = \begin{bmatrix} 3 & 2 \\ 2 & 3 \end{bmatrix}$

3. $\begin{bmatrix} -2 & -5 \\ -3 & 0 \end{bmatrix}$

5. $\begin{bmatrix} 2 & 3 & 4 \\ 0 & 0 & 0 \\ 4 & 6 & 8 \end{bmatrix}$

7. $\begin{bmatrix} \frac{7}{6} & \frac{1}{3} & 2 \\ \frac{2}{3} & \frac{1}{2} & \frac{5}{6} \end{bmatrix}$

9. $\begin{bmatrix} 1 & 0 & 0 \\ 0 & 4 & 0 \\ 0 & 0 & 9 \end{bmatrix}$

11. $\begin{bmatrix} 0 & 0 & 0 \\ 0 & 0 & 0 \\ 3 & 0 & 0 \end{bmatrix}$

13. $X = \begin{bmatrix} -\frac{1}{5} \\ \frac{2}{5} \end{bmatrix}$

15. $X = \begin{bmatrix} \frac{3}{7} \\ \frac{2}{7} \end{bmatrix}$

17. $X = \begin{bmatrix} 0 \\ -1 \end{bmatrix}$

19. $\begin{bmatrix} 1 & 1 & 1 \\ 1 & 0 & 0 \\ 1 & 1 & 0 \end{bmatrix}$

21. $\begin{bmatrix} 1 & 2 & 2 & 2 \\ 1 & 0 & 0 & 0 \\ 0 & 0 & 0 & 0 \\ 1 & 2 & 0 & 0 \end{bmatrix}$

Section 6.3

1. $x = -1$
   $y = 0$
   $z = 1$

3. $x = -\frac{1}{4}$
   $y = 0$
   $z = \frac{1}{4}$

5. $x = \frac{7}{16}$
   $y = \frac{5}{16}$
   $z = \frac{1}{16}$

7. $x = 2$
   $y = -1$
   $z = -6$

Section 6.4

1. $\begin{bmatrix} -1 & 0 \\ -1 & -1 \end{bmatrix}$

3. $\begin{bmatrix} -\frac{2}{17} & -\frac{6}{17} \\ \frac{6}{17} & \frac{1}{17} \end{bmatrix}$

5. No inverse exists.

7. $\begin{bmatrix} \frac{7}{4} & -\frac{1}{4} & -\frac{9}{4} \\ 1 & 0 & -1 \\ -\frac{7}{4} & \frac{1}{4} & \frac{5}{4} \end{bmatrix}$

9. $\begin{bmatrix} \frac{5}{16} & \frac{7}{16} & \frac{3}{13} \\ -\frac{1}{16} & \frac{5}{16} & \frac{9}{16} \\ \frac{3}{16} & \frac{1}{16} & \frac{5}{16} \end{bmatrix}$

11. $\begin{bmatrix} \frac{1}{3} & 0 & 0 \\ 0 & \frac{1}{3} & 0 \\ 0 & 0 & \frac{1}{3} \end{bmatrix}$

13. $\begin{bmatrix} 1 & 1 & 0 & -1 \\ -1 & -3 & -1 & 2 \\ 0 & -1 & 0 & 1 \\ 1 & 2 & 1 & -2 \end{bmatrix}$

15. Carry out the multiplication $A \cdot A^{-1}$.

17. Carry out the multiplication $A \cdot A^{-1}$.

Section 6.5

1. The system

$$x + y + z + t = 16$$
$$7x + 8y + 6z + 4t = 60$$
$$4x + 2y + 4z + 2t = 30$$

reduces to the system

$$x - 4t = -70$$
$$y + t = 17$$
$$z + 4t = 69$$

giving

$$x = -70 + 4t \geq 0$$
$$y = \quad 17 - \quad t \geq 0$$
$$z = \quad 69 - 4t \geq 0$$

This gives the incompatible inequalities $t \geq \frac{35}{2} = 17.5$ and $t \leq 17$, and hence the answer is No.

3. $x = -1$ 5. $x = 3$ 7. $x = \quad 1$ 9. $x = 3$
      $y = \quad 1$ $y = 1$ $y = -1$ $y = 1$
      $z = \quad 4$ $z = 5$ $z = \quad 2$ $z = 5$

11. $x = \quad 2$ 13. $x = -2z + 6$
       $y = -2t - 2$ $y = \quad z - 6$
       $z = \quad 3t + 1$

Section 6.6

1.

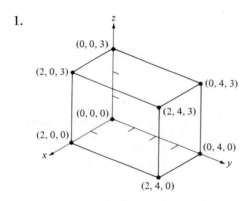

3.   The plane that is parallel to the xz-plane and passes through the point $(0, 3, 0)$ on the y-axis

CHAPTER 7

Section 7.1

1.   (a)   The probability is q after one transition, $p \times q \times q = pq^2$ after three transitions.

     (b)   The possible chains of transitions are $a_2 \xrightarrow{p} a_3 \xrightarrow{p} a_4 \xrightarrow{q} a_3 \xrightarrow{q} a_2$ and $a_2 \xrightarrow{p} a_3 \xrightarrow{q} a_2 \xrightarrow{p} a_3 \xrightarrow{q} a_2$ and the probability is $p^2q^2 + p^2q^2 = 2p^2q^2 = 2(pq)^2$. Taking $p = \frac{1}{4}$ and $q = \frac{3}{4}$ gives $2(\frac{3}{16})^2 = \frac{9}{128} \approx 0.070$.

3.   (a)   The probability is 0, because an even number of transitions is required.

(b)   We start and end in state a_3. The possible chains of transition are

$$a_3 \rightarrow a_2 \rightarrow a_3 \rightarrow a_2 \rightarrow a_3$$
$$a_3 \rightarrow a_4 \rightarrow a_3 \rightarrow a_4 \rightarrow a_3$$
$$a_3 \rightarrow a_2 \rightarrow a_3 \rightarrow a_4 \rightarrow a_3$$
$$a_3 \rightarrow a_4 \rightarrow a_3 \rightarrow a_2 \rightarrow a_3$$

Because $p = q = \frac{1}{2}$, each chain has the probability $\frac{1}{2} \times \frac{1}{2} \times \frac{1}{2} \times \frac{1}{2} = \frac{1}{2^4}$; the answer is $4 \times \frac{1}{2^4} = \frac{1}{4}$.

5.

7.

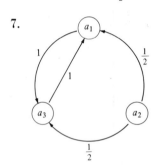

9.

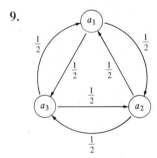

11.

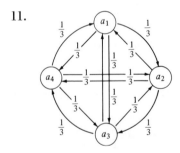

13.

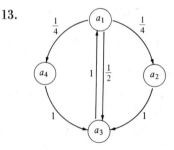

15.

	a_1	a_2	a_3
a_1	0	$\frac{1}{2}$	$\frac{1}{2}$
a_2	0	$\frac{1}{2}$	$\frac{1}{2}$
a_3	1	0	0

17.

	a_1	a_2	a_3	a_4
a_1	$\frac{1}{5}$	$\frac{2}{5}$	0	$\frac{2}{5}$
a_2	0	0	$\frac{1}{2}$	$\frac{1}{2}$
a_3	0	1	0	0
a_4	0	1	0	0

Section 7.2

1.   (b), (c), (d), (f), and (h)

3.   (a) $p^{(1)} = \begin{bmatrix} \frac{2}{3} & \frac{1}{3} \end{bmatrix}$ $p^{(2)} = \begin{bmatrix} \frac{4}{9} & \frac{5}{9} \end{bmatrix}$ $p^{(3)} = \begin{bmatrix} \frac{14}{27} & \frac{13}{27} \end{bmatrix}$

     (b) $p^{(1)} = \begin{bmatrix} \frac{4}{9} & \frac{5}{9} \end{bmatrix}$ $p^{(2)} = \begin{bmatrix} \frac{14}{27} & \frac{13}{27} \end{bmatrix}$ $p^{(3)} = \begin{bmatrix} \frac{40}{81} & \frac{41}{81} \end{bmatrix}$

(c) $p^{(1)} = [\frac{1}{3} \quad \frac{2}{3}]$ $p^{(2)} = [\frac{5}{9} \quad \frac{4}{9}]$ $p^{(3)} = [\frac{13}{27} \quad \frac{14}{27}]$

(d) $p^{(1)} = [\frac{7}{12} \quad \frac{5}{12}]$ $p^{(2)} = [\frac{17}{36} \quad \frac{19}{36}]$ $p^{(3)} = [\frac{55}{108} \quad \frac{53}{108}]$

5. $T^2 = \begin{bmatrix} \frac{3}{8} & \frac{5}{8} \\ \frac{5}{16} & \frac{11}{16} \end{bmatrix}$ $T^3 = \begin{bmatrix} \frac{11}{32} & \frac{21}{32} \\ \frac{21}{64} & \frac{43}{64} \end{bmatrix}$

7. $T^2 = \begin{bmatrix} 1 & 0 \\ 0 & 1 \end{bmatrix}$ $T^3 = \begin{bmatrix} 0 & 1 \\ 1 & 0 \end{bmatrix}$

9. $T^2 = \begin{bmatrix} 0 & \frac{2}{3} & \frac{1}{3} \\ 0 & 1 & 0 \\ 0 & 0 & 1 \end{bmatrix}$ $T^3 = T$

11. $T^2 = \begin{bmatrix} \frac{1}{2} & \frac{1}{4} & \frac{1}{4} \\ \frac{1}{4} & \frac{1}{2} & \frac{1}{4} \\ \frac{1}{4} & \frac{1}{4} & \frac{1}{2} \end{bmatrix}$ $T^3 = \begin{bmatrix} \frac{3}{8} & \frac{3}{8} & \frac{1}{4} \\ \frac{3}{8} & \frac{1}{4} & \frac{3}{8} \\ \frac{1}{4} & \frac{3}{8} & \frac{3}{8} \end{bmatrix}$

13. $T^2 = \begin{bmatrix} \frac{3}{8} & \frac{5}{16} & \frac{5}{16} \\ \frac{5}{16} & \frac{3}{8} & \frac{5}{16} \\ \frac{5}{16} & \frac{5}{16} & \frac{3}{8} \end{bmatrix}$ $T^3 = \begin{bmatrix} \frac{21}{64} & \frac{21}{64} & \frac{11}{32} \\ \frac{21}{64} & \frac{11}{32} & \frac{21}{64} \\ \frac{11}{32} & \frac{21}{64} & \frac{21}{64} \end{bmatrix}$

15. $T^2 = \begin{bmatrix} 1 & 0 & 0 & 0 \\ \frac{1}{2} & \frac{1}{4} & 0 & \frac{1}{4} \\ \frac{1}{4} & 0 & \frac{1}{4} & \frac{1}{2} \\ 0 & 0 & 0 & 1 \end{bmatrix}$ $T^3 = \begin{bmatrix} 1 & 0 & 0 & 0 \\ \frac{5}{8} & 0 & \frac{1}{8} & \frac{1}{4} \\ \frac{1}{4} & \frac{1}{8} & 0 & \frac{5}{8} \\ 0 & 0 & 0 & 1 \end{bmatrix}$

17. $T^2 = \begin{bmatrix} 0 & 0 & \frac{1}{2} & \frac{1}{2} \\ \frac{1}{4} & \frac{1}{4} & \frac{1}{4} & \frac{1}{4} \\ 0 & 0 & 1 & 0 \\ 0 & 0 & 0 & 1 \end{bmatrix}$ $T^3 = \begin{bmatrix} 0 & 0 & \frac{1}{2} & \frac{1}{2} \\ \frac{1}{8} & \frac{1}{8} & \frac{3}{8} & \frac{3}{8} \\ 0 & 0 & 1 & 0 \\ 0 & 0 & 0 & 1 \end{bmatrix}$

19. $p^{(1)} = [0 \quad 0 \quad 1]$ $p^{(2)} = [0 \quad 1 \quad 0]$ $p^{(3)} = [0 \quad 0 \quad 1]$

   $p^{(1)} = [0 \quad \frac{1}{3} \quad \frac{2}{3}]$ $p^{(2)} = [0 \quad \frac{2}{3} \quad \frac{1}{3}]$ $p^{(3)} = [0 \quad \frac{1}{3} \quad \frac{2}{3}]$

Section 7.3

1. (a), (c), and (f) are regular; (b), (d), (e), (g), and (h) are not regular.

3. Follow Example 4 with $[u_1 \quad u_2] = [u_1 \quad u_2] \begin{bmatrix} \frac{1}{2} & \frac{1}{2} \\ \frac{1}{2} & \frac{1}{2} \end{bmatrix}$. This will give the system of equations $\frac{1}{2}u_1 + \frac{1}{2}u_2 = u_1$, $\frac{1}{2}u_1 + \frac{1}{2}u_2 = u_2$, $u_1 + u_2 = 1$, and these give the probability vector $u = [\frac{1}{2} \quad \frac{1}{2}]$.

5. Follow Example 4 with $S = \begin{array}{c} \\ C \\ W \end{array}\begin{array}{c} C \quad W \\ \begin{bmatrix} \frac{9}{10} & \frac{1}{10} \\ \frac{24}{25} & \frac{1}{25} \end{bmatrix} \end{array}$. The resulting probability vector is $u = [\frac{48}{53} \quad \frac{5}{53}]$.

7. The transition matrix is

$T = \begin{array}{c} \\ (a) \\ (b) \\ (c) \end{array}\begin{array}{ccc} (a) & (b) & (c) \\ \begin{bmatrix} \frac{1}{2} & \frac{1}{4} & \frac{1}{4} \\ \frac{1}{4} & \frac{1}{2} & \frac{1}{4} \\ 0 & \frac{1}{2} & \frac{1}{2} \end{bmatrix} \end{array}$

(a) This is a regular process because all the elements of T^2 are positive.

(b) The probability vector is $u = [\frac{2}{9} \quad \frac{4}{9} \quad \frac{1}{3}]$, and so the probability we seek is $\frac{2}{9}$.

Section 7.4

1. not absorbing

3. absorbing, with absorbing state a_3

$$\begin{array}{c} & \begin{array}{ccc} a_3 & a_2 & a_1 \end{array} \\ \begin{array}{c} a_3 \\ a_2 \\ a_1 \end{array} & \left[\begin{array}{c|cc} 1 & 0 & 0 \\ \hline \frac{1}{3} & \frac{1}{3} & \frac{1}{3} \\ 1 & 0 & 0 \end{array}\right] \end{array}$$

5. not absorbing
7. not absorbing

9. absorbing, with absorbing state a_2 and a_3

$$\begin{array}{c} & \begin{array}{cccc} a_2 & a_3 & a_1 & a_4 \end{array} \\ \begin{array}{c} a_2 \\ a_3 \\ a_1 \\ a_4 \end{array} & \left[\begin{array}{cc|cc} 1 & 0 & 0 & 0 \\ 0 & 1 & 0 & 0 \\ \hline 0 & \frac{1}{2} & \frac{1}{2} & 0 \\ \frac{1}{4} & \frac{1}{4} & \frac{1}{4} & \frac{1}{4} \end{array}\right] \end{array}$$

11. absorbing, with absorbing state a_4 and a_5

$$\begin{array}{c} & \begin{array}{ccccc} a_4 & a_5 & a_1 & a_2 & a_3 \end{array} \\ \begin{array}{c} a_4 \\ a_5 \\ a_1 \\ a_2 \\ a_3 \end{array} & \left[\begin{array}{cc|ccc} 1 & 0 & 0 & 0 & 0 \\ 0 & 1 & 0 & 0 & 0 \\ \hline 0 & \frac{1}{7} & \frac{2}{7} & \frac{2}{7} & \frac{2}{7} \\ \frac{2}{7} & \frac{2}{7} & \frac{1}{7} & \frac{2}{7} & 0 \\ \frac{2}{7} & \frac{2}{7} & \frac{2}{7} & 0 & \frac{1}{7} \end{array}\right] \end{array}$$

13. $I - H = \begin{bmatrix} \frac{3}{4} & -\frac{1}{4} \\ -\frac{1}{4} & \frac{3}{4} \end{bmatrix}$
$(I - H)^{-1} = \begin{array}{c} & \begin{array}{cc} a_3 & a_4 \end{array} \\ \begin{array}{c} a_3 \\ a_4 \end{array} & \begin{bmatrix} \frac{3}{2} & \frac{1}{2} \\ \frac{1}{2} & \frac{3}{2} \end{bmatrix} \end{array}$

$$N \cdot F = \begin{bmatrix} \frac{3}{2} & \frac{1}{2} \\ \frac{1}{2} & \frac{3}{2} \end{bmatrix} \cdot \begin{bmatrix} \frac{1}{2} & 0 \\ 0 & \frac{1}{2} \end{bmatrix} = \begin{array}{c} & \begin{array}{cc} a_1 & a_2 \end{array} \\ \begin{array}{c} a_3 \\ a_4 \end{array} & \begin{bmatrix} \frac{3}{4} & \frac{1}{4} \\ \frac{1}{4} & \frac{3}{4} \end{bmatrix} \end{array}$$

(a) $\frac{1}{2}$

(b) 2

(c) $\frac{1}{4}; \frac{3}{4}$

INDEX

INDEX